Quantitative Methods in the Humanities and Social Sciences

Quantitative Methods in the Humanities and Social Sciences is a book series designed to foster research-based conversation with all parts of the university campus – from buildings of ivy-covered stone to technologically savvy walls of glass. Scholarship from international researchers and the esteemed editorial board represents the far-reaching applications of computational analysis, statistical models, computer-based programs, and other quantitative methods. Methods are integrated in a dialogue that is sensitive to the broader context of humanistic study and social science research. Scholars, including among others historians, archaeologists, new media specialists, classicists and linguists, promote this interdisciplinary approach. These texts teach new methodological approaches for contemporary research. Each volume exposes readers to a particular research method. Researchers and students then benefit from exposure to subtleties of the larger project or corpus of work in which the quantitative methods come to fruition.

Harald Klinke

Cultural Data Science

An Introduction to R

Springer

Harald Klinke
Department of Art History
Ludwig Maximilian University of Munich (LMU)
München, Germany

ISSN 2199-0956 ISSN 2199-0964 (electronic)
Quantitative Methods in the Humanities and Social Sciences
ISBN 978-3-031-88129-9 ISBN 978-3-031-88130-5 (eBook)
https://doi.org/10.1007/978-3-031-88130-5

This Springer imprint is published by the registered company Springer Nature Switzerland AG
The registered company address is: Gewerbestrasse 11, 6330 Cham, Switzerland

For Liska

Preface

The intersection between humanities, media, and cultural studies and computational methods has become one of the most dynamic areas of research and teaching in universities today. However, as this interdisciplinary paradigm grows, many students and researchers find themselves unprepared for the technical demands of coding and data analysis. *Cultural Data Science: An Introduction to R* addresses this challenge with clarity and approachability. Designed as a beginner-friendly resource, this book equips students with the fundamental skills to work with data while grounding their learning in culturally relevant examples from museum collections and art historical archives. Such a low-barrier introduction is precisely what is needed to make computational approaches accessible to a wider audience, empowering the next generation of scholars to bridge the gap between cultural analysis and data analysis.

Harald Klinke is a pioneer in digital art history and cultural data science, having contributed significantly to the field through his research, teaching, and role as co-founder of the *International Journal for Digital Art History*. Through his work, he demonstrates that data science is not just a technical tool but a fundamental approach to understanding cultural phenomena in today's data-rich environment.

This book makes an important contribution by recognizing the need for a resource that demystifies cultural data analysis while addressing the specific needs of humanities and cultural data science students. By focusing on R, a powerful yet accessible programming language, and using real-world examples from major museum collections, this book offers a clear path forward for those eager to engage with data analysis methods but unsure where to start. I am delighted to see this book come to fruition and confident that it will become a cornerstone resource for educators and students alike in the growing fields of cultural data science and cultural analytics.

The release of this textbook represents an important step in the democratization of cultural analysis. When I first began developing cultural analytics methods in the mid-2000s, there was no software suitable for working with large cultural image collections, and museums had not yet started making their collections available online. Today, through open-source tools like R and freely available cultural datasets from many major museums, we are witnessing a transformation in who can participate in computational analysis of culture. This book accelerates this democratization by providing a clear entry point for humanities and media scholars and students to develop the data science literacy needed in our software society. By making these tools and methods accessible to a broader audience, we expand the possibilities for new insights and perspectives in cultural research.

One of the most exciting aspects of computational analysis of cultural data is its ability to reveal patterns and relationships that remain invisible to traditional methods of study. When working with museum collections that contain hundreds of thousands of artifacts or analyzing millions of contemporary digital images, the human eye and mind alone cannot grasp the full scope of visual and temporal patterns. The methods taught in this book—from basic statistical analysis to specific visualization techniques suitable for working with image collections—enable scholars to zoom out to see broad historical patterns while still maintaining the ability to zoom in on specific artifacts and their contexts. For example, using the tools and approaches presented here, students can trace the evolution of visual styles across decades, analyze the changing dynamics of cultural institutions through their acquisition patterns, or map complex networks of artistic influence—all while maintaining the methodological rigor that cultural research demands.

The rise of computational analysis of visual media comes at a particularly significant moment in cultural history. For the first time, we can systematically analyze both historical artistic traditions and the massive scale of contemporary visual culture—from museum collections to social media and AI imagery. This book introduces students to the essential tools for this analysis, showing how computational methods can help us understand both historical visual artifacts and the new forms of digital visual culture that increasingly shape our world. By combining traditional art historical knowledge with computational approaches, students can develop new frameworks for analyzing visual culture across historical periods and media forms.

Looking toward the future, this book arrives at a crucial moment in the evolution of digital humanities and cultural analytics. We are moving beyond the initial excitement of digital methods toward more sophisticated hybrid approaches that combine computational analysis with traditional humanities methodologies. The examples and methods presented here lay groundwork for students to develop innovative analytical frameworks that bridge quantitative and qualitative approaches. As cultural institutions continue to digitize their collections and as new computational tools emerge, the fundamental skills taught in this book will become increasingly valuable. The next generation of scholars will need to move fluidly between close and distant reading, between individual artifacts and large-scale patterns, and between human interpretation and computational analysis. This book provides an essential foundation for developing these hybrid competencies that will shape the future of cultural research.

Graduate Center
The Graduate Center, CUNY
New York, USA

Lev Manovich

Acknowledgements

Munich, May 2025

I would like to express my deepest gratitude to everyone who has supported and inspired me on this journey.

I am deeply grateful to my colleagues and students of Ludwig-Maximilians-Universität München, FernUniversität Hagen, and the participants and organizers of the Digital Art History Summer School in Málaga for their inspiring engagement in numerous seminars and workshops. Their curiosity, enthusiasm, and thought-provoking contributions have shaped the foundation of this book.

A heartfelt thanks to Eleonora for her meticulous proofreading of the manuscript, ensuring clarity and precision throughout. I am also indebted to the team at Springer Press for their professional support and guidance during the creation of this book.

I extend my sincere gratitude to Lev Manovich, whose groundbreaking work and enriching conversations have been a constant source of inspiration over the years. Most importantly, I would like to thank Liska, whose unwavering support, encouragement, and companionship have been the cornerstone of this journey—and in all my endeavors.

Harald Klinke

Competing Interests The author has no competing interests to declare that are relevant to the content of this manuscript.

Contents

Chapter 1
Introduction

Today, the amount of available data is steadily increasing across all disciplines, opening up new avenues for research. As artificial intelligence enters the humanities, the question of how cultural studies can meaningfully harness this abundance of data is moving into the spotlight. In the context of exploring the role of data science within the discipline of art history, it is essential to begin by defining the very essence of data science itself. Data science is a broad and dynamic field that can be approached from different perspectives, each shedding light on its unique facets.

One prominent definition characterizes data science as an interdisciplinary field that uses scientific methods, processes, algorithms, and systems to extract knowledge and insights from structured and unstructured data [1]. In this view, data science serves as a bridge between raw data and actionable knowledge. It encompasses the skill of transforming data into meaningful insights with the aid of information technology. This perspective aligns with the conventional notion of data science as a discipline focused on turning data into knowledge.

Another definition of data science frames it as a concept that utilizes statistics, data analysis, machine learning, and domain knowledge to comprehend real-world phenomena from data [2]. Here, data science is inherently interdisciplinary, drawing from the realms of statistics, data analysis, machine learning, and subject-specific domain knowledge. The primary goal is to gain a profound understanding of tangible phenomena, such as art or the history of art, by making use of data.

Figure 1.1 shows this perspective as a data-driven approach to solving complex questions: On the left side, we see the realm of computer science, where data infrastructure, data preprocessing, and related technologies thrive. On the right side lies the domain of statistics, encompassing essential statistical knowledge and methodological applications.

In the dynamic space between these two realms, we find programming languages and machine learning techniques, serving as tools for extracting insights from data. In this intermediary space, the application of computational methodologies meets the world of statistics.

H. Klinke, *Cultural Data Science*, Quantitative Methods in the Humanities and Social Sciences, https://doi.org/10.1007/978-3-031-88130-5_1

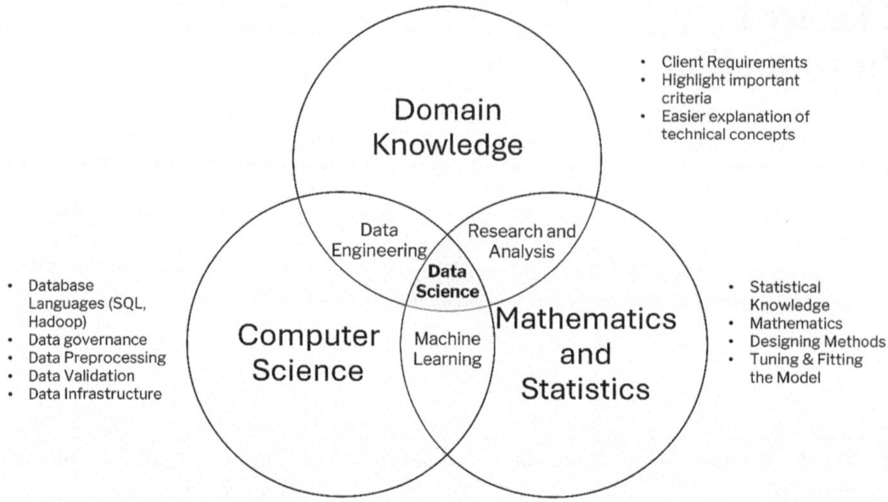

Fig. 1.1 Data Science as an interdisciplinary field

However, a third element is necessary: the domain knowledge. Whether it is art history, economics, or any other field, this domain knowledge guides the data science process, ensuring that the insights derived are both contextually relevant and meaningful. Thus, data science acts as a conduit between data and knowledge, bridging the gap between raw information and actionable insights. Its multifaceted nature makes data science a powerful tool for unraveling complex questions and gaining a deeper understanding of the world around us. In the field of cultural data science within the particular context of art history, these diverse perspectives on data science will play a pivotal role in shaping our approach and enriching our insights.

The landscape of cultural research is undergoing a profound transformation, finding itself intricately woven into the interdisciplinary fabric of the broader project known as data science. It is within this vast arena of data science that cultural research discovers fertile ground for insights and innovation.

Digital art history emerges as a key player in this interdisciplinary landscape. It not only brings its wealth of domain-specific knowledge to the table but also draws upon the rich tapestry of computer science and statistics. In this convergence, the boundaries between disciplines blur, and a new realm of opportunities emerges.

It is within this collaborative, interdisciplinary approach that cultural data science finds its true meaning. Art history remains a distinct field, with its unique historically developed methodologies and practices. However, it now expands its horizons, reaching across disciplinary borders to embrace the tenets of data science.

So, what exactly is "cultural data science" in this broader context, or perhaps more aptly, what can we refer to as "data science and art history"? Traditionally, art historians gravitate toward institutions like museums. They meticulously examine individual artworks, such as paintings, and engage in nuanced comparisons, drawing upon their mental archives of prior observations. Extensive reading and research further characterize their approach.

However, the advent of cultural data science ushers in a new paradigm. This paradigm recognizes that art history, while preserving its core principles, is enriched by data-driven methodologies and computational techniques. It acknowledges that the analysis of cultural artifacts benefits from the systematic examination of large-scale data, opening doors to fresh perspectives and insights.

Cultural data science represents a fusion of disciplines. It retains the essence of art history while embracing the tools and practices of data science. It is an exciting journey that holds the promise of unveiling hidden narratives, unlocking the potential of technology, and enriching our understanding of cultural phenomena. As we delve deeper into cultural data science, we witness its transformative power to art history in the digital age.

While the conventional approach to art history often involves close examination of individual artworks and meticulous research, cultural data science introduces a powerful extension to these established methodologies. The Museum of Modern Art (MoMA), for instance, offers an intriguing opportunity for exploration. It generously provides access to data encompassing its vast collection, comprising an impressive 140,000 objects, available for download on the online repository GitHub [3].

With this treasure trove of data at our disposal, we can embark on a journey that goes beyond traditional art historical analysis. Here is a glimpse of what becomes possible:

1. Visualizing Collection Evolution: By working with the MoMA‘s extensive dataset, we can trace the growth and transformation of its collection over time. This allows us to discern shifts in artistic signatures and the evolving impact of movements like modernism on the institution. Moreover, it enables us to compare MoMA‘s collection with that of other museums, shedding light on their respective collection policies and trajectories.
2. Harnessing Image Analysis: Beyond textual data, images hold a wealth of information. Through computer vision and image analysis, we can uncover hidden patterns and insights within the visual realm of art. This may involve categorizing artworks based on visual characteristics, identifying recurring themes or styles, and much more.
3. Interactive Visualizations: Data science empowers us to create dynamic, interactive visualizations. We can navigate and explore collections intuitively, zooming in on details, grouping artworks by similarity, or filtering by specific keywords. For example, we can investigate depictions of women in art, juxtaposing them with other thematic elements or exploring their representation throughout history.

These examples merely scratch the surface of what is achievable within the realm of cultural data science. The toolkit provided by data science and statistics is extensive, encompassing image analysis, natural language processing, network visualization, timeline analysis, similarity assessments, maybe even predictive analytics.

In practice, we have witnessed the application of these methods to substantial datasets in art history. Machine learning aids in recognizing artistic styles and attributions, network analysis unveils hidden connections within the art world, and predictive analytics holds potential in art market analysis [4].

Cultural data science is not an exclusive realm reserved for experts. It is an accessible field that welcomes newcomers. For years, I have been teaching students the power of data science predominantly at the LMU Munich and the Digital Art History Summer School at the University of Málaga, equipping them with the tools, statistics, and datasets necessary to embark on their own projects. The results of their efforts, often undertaken with no prior programming knowledge, can be explored on my website.

> ℹ Note: You can find further and updated information to this book on my website https://www.harald-klinke.de/cultural-data-science. It includes all the code from this book.

These students have tapped into the transformative potential of cultural data science, using programming languages, statistics, and rich datasets from institutions like MoMA and the Rijksmuseum. The possibilities are vast, and the outcomes illustrate the impact of data science on cultural research.

What makes data science truly fascinating is the structured process it follows—a journey often referred to as the data science pipeline. While it might seem like a data-driven adventure, it is fundamentally rooted in questions. It is more aptly described as a "question-driven process" because it always commences with a pivotal inquiry: What is our research goal? What knowledge are we seeking, and which data will guide our path? Are those data readily available, or do we need to procure or generate them?

Once we have acquired the data, the journey takes us through a process of exploration. We need to gain an understanding of our data, whether it involves obtaining an overview, extracting relevant subsets, or uncovering latent questions that lie within the dataset's depths.

As indicated by the arrows in the diagram (Fig. 1.2), the data science process is inherently iterative. The most crucial iterations occur when our initial overview of the data eventually sparks new questions. Ultimately, we might unearth unexpected answers that lead us in novel directions. In the end, we have results, visualizations, and more, which must be tethered back to our original questions. We refine and reformulate our inquiries as our understanding deepens, allowing us to craft a narrative—one that redefines our perception of the history of art.

This process starts with qualitative questions and qualitative methods, aligning with the practice of art history. However, we introduce quantitative methods and computational thinking as we delve into the realm of data science. This juncture, represented by the blue segment in Fig. 1.2, embodies "computational thinking." Here, we dissect complex tasks into manageable subtasks, translating them into a

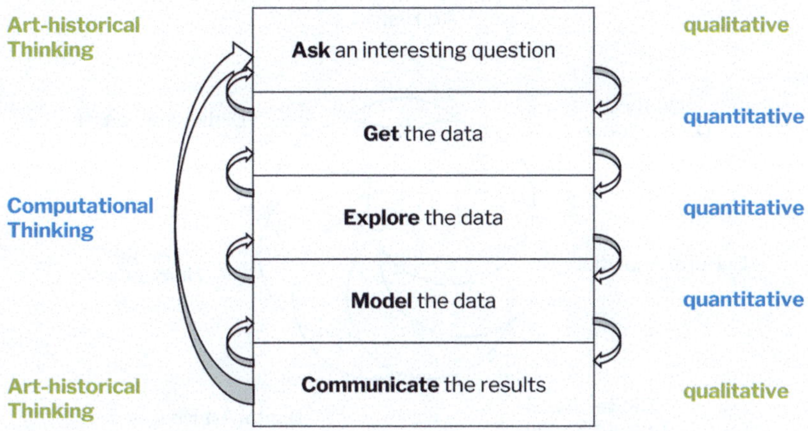

Fig. 1.2 The data science process [5]

programming language, and iteratively optimizing them [6]. It is a realm where we harness the power of data and computational tools.

However, it is crucial not to lose sight of our roots in art historical thinking. We must remain grounded in our research questions and objectives, ensuring that data science does not solely dictate our narrative. It is about a symbiotic relationship between computational thinking and humanities thinking. As scholars and practitioners in our own field, we must impart computational thinking to our students, equipping them with the skills to navigate the data-driven landscape. In parallel, information science should embrace "humanities thinking" to effectively bridge the gap between data-driven methodologies and the nuanced research questions that define the humanities.

This interdisciplinary collaboration is where the true potential of cultural data science comes to life, enabling us to explore art history's uncharted territories, discover hidden narratives, and redefine the boundaries of our field.

So, what do we call this emerging field? There is the well-established discipline of art history, and we have witnessed the birth of fields like "Bildwissenschaft" in German-speaking countries or "Visual Studies" in others, which extend our purview beyond solely artistic images. This expansion naturally prompts questions about whether it is still art history or something new altogether—an expansion.

Then there are digital humanities: an interdisciplinary field that combines the use of digital tools and methods with traditional humanities research to analyze and interpret cultural, historical, and literary data. Digital art history could be considered a subfield of digital humanities. Yet, it carries a distinct perspective—seeking to introduce quantitative methods to a humanities field deeply rooted in visual interpretation.

As shown in Fig. 1.3, we could also consider terms like "digital visual culture studies," which aligns with "Bildwissenschaft." Another possibility is "cultural analytics." A term introduced by Lev Manovich who, in his early contributions to

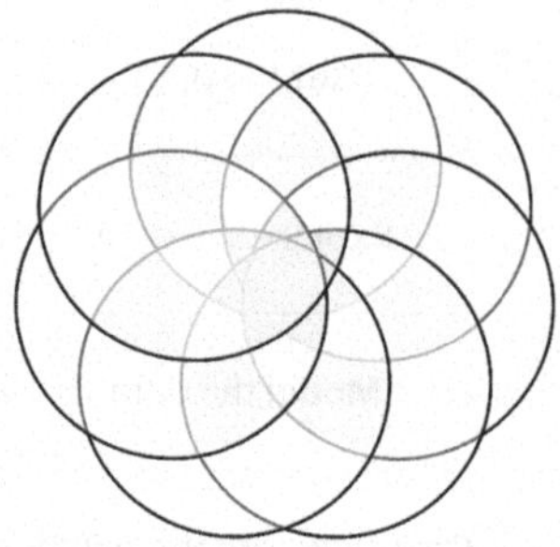

Fig. 1.3 Terms related to Cultural Data Science

International Journal for Digital Art History, touched upon a realm where methods are applicable to cultural phenomena, whether they are artworks or other facets of culture [7]. I propose the term "cultural data science" to highlight the specific application of data science methods to cultural research questions, emphasizing the intersection between computational techniques and the analysis of cultural phenomena. While the words may differ, the essence remains the same.

No matter what label we affix, I prefer to think of it as "art history with new methods" or simply "data science for art history."

So, how can we define cultural data science? It is the fusion of qualitative and quantitative methods, as previously explained—a data-driven art history with a central question: How can computers enhance our comprehension of the historical and contemporary developments in art? These questions have always been intrinsic to our discipline; we are now exploring them with innovative tools and methodologies.

Throughout the history of science, there have been distinct paradigms, primarily observed in the natural sciences. Initially, the sciences were empirical, focusing on observing and describing natural phenomena. This empirical approach was foundational, serving as the building block for subsequent paradigms.

The second paradigm expanded on empirical observations by introducing models. In the natural sciences, these models often took the form of mathematical equations and formulas. They allowed scientists to not only *describe* phenomena but also *predict* how they would behave under different conditions. With the advent of computers, these models evolved into powerful simulation tools, enabling researchers to conduct experiments in virtual environments that might be impractical or impossible in reality—think of complex climate models, for instance.

However, the most recent paradigm shift is where data-driven sciences take center stage. This shift is not limited to the natural sciences; it permeates all disciplines, including the humanities. Jim Gray, an influential information scientist, aptly stated that "everything about science is changing because of the impact of information technology." [8]

Data-driven sciences leverage unprecedented access to vast volumes of data and computational capabilities to process, analyze, and derive knowledge from this data. This paradigm shift has fundamentally altered the way research is conducted across various fields, including art history. It has propelled us into an era where data science plays a pivotal role in reshaping the landscape of cultural research (Table 1.1).

The traditional notion of art history is evolving as the digital era offers new possibilities for data-driven research. With structured, machine-readable, multimodal, and large datasets at our disposal, data science empowers us to venture beyond the confines of conventional art history. It marks a shift from traditional manual methods to a data-driven approach that can harness the power of computational analysis. We can now explore the long tail of art history, examining obscure or overlooked details. Through data analysis tools, we can automatically or semi-automatically cluster datasets, uncover similarities, delve into network structures, and engage in both close and distant viewing. This allows us to grasp the entirety of a dataset while zooming in on individual objects within their broader context.

The goal of this data-driven paradigm is to recognize patterns, distill valuable insights, and contextualize individual objects within the vast sea of data. It ushers in a new foundation for art historical research, revolutionizing the way we perceive, study, and interpret the cultural artifacts that shape our understanding of human history. Nevertheless, it is essential to acknowledge the limitations, such as data biases, the insufficiency of data for certain research questions, and the need for human interpretation in data analysis.

In conclusion, data science enriches art history by expanding our analytical toolkit and providing fresh perspectives. It is not only about harnessing powerful tools but also about adopting ethical considerations and critical thinking. This enrichment does not replace traditional methods but complements them, offering new computational methods that unlock insights and facilitate the analysis of extensive datasets. The blend of qualitative and quantitative methods presents a compelling

Table 1.1 Paradigms of science [9]

1st paradigm: Empirical science	**2nd paradigm:** Model-based theoretical science	**3rd paradigm:** Computational science (simulations)	**4th paradigm:** (Big) data driven science
Experiments	Laws of Thermodynamics	Density Functional Theory, Molecular Dynamics	Predictive analytics, Clustering, Relationship mining, Anomaly detection

1600 1950 2000

opportunity for art history, empowering scholars to address complex questions with more nuanced perspectives to understand our cultural history.

The raison d'être of this book lies in its mission to empower cultural researchers and enthusiasts from various disciplines to actively participate in the interdisciplinary realm of data science. To engage in meaningful conversations with information science experts and unlock the vast potential that data science methods offer for cultural research, it is imperative to become proficient in these methods.

Drawing from years of experience teaching cultural data science at LMU Munich, this book emerges as a valuable resource for those who may not have access to formal workshops or courses but yearn for a comprehensive guide on how to navigate the intricate landscape of cultural data. Many students and researchers have encountered situations where they possess valuable cultural data, often stored in Excel sheets, yet face the daunting challenge of analyzing this data to address their specific research questions. This book serves as an introduction, offering guidance for individuals at all levels, even and especially for those starting with no prior knowledge in the field.

The purpose of this book is to democratize the knowledge and skills required to work with cultural data and digital methods effectively. It is not just a domain for the select few, but a vibrant and inclusive space where everyone can contribute to our collective understanding of the world's rich cultural heritage. Whether you are a seasoned researcher or a curious novice, this book is designed to equip you with the tools and techniques needed to gain new insights into our cultural past. It is within the narratives of our past that we find the threads that define our present and future.

It is my hope that this textbook will serve as a companion for those who seek to unearth the hidden treasures within cultural data, revealing the stories that shape our shared human experience.

References

1. What is Data Science? Definition, Examples, Tools & More|DataCamp. 2024. https://www.datacamp.com/blog/what-is-data-science-the-definitive-guide.
2. Hayashi C. What is data science? Fundamental concepts and a heuristic example. In: Hayashi C, Yajima K, Bock HH, Ohsumi N, Tanaka Y, Baba (Hg) Y, editors. Data Science, Classification, and Related Methods. Proceedings of the Fifth Conference of the International Federation of Classification Societies (IFCS-96), Kobe, Japan, March 27–30, 1996. Tokyo: Springer Japan (Studies in Classification, Data Analysis, and Knowledge Organization); 1998. p. 40–51.
3. GitHubMuseumofModernArt/collection: The Museum of Modern Art (MoMA) collection data. 2024. https://github.com/MuseumofModernArt/collection.
4. Many examples you will find in the International Journal for Digital Art History: https://dahj.org/articles.
5. This diagram is based on the Harvard course CS 109 by Joe Blitzstein and Hanspeter Pfister: Quora. 2024. What is it like to design a data science class?. https://www.quora.com/What-is-it-like-to-design-a-data-science-class-In-particular-what-was-it-like-to-design-Harvards-

new-data-science-class-taught-by-professors-Joe-Blitzstein-and-Hanspeter-Pfister/answer/Joe-Blitzstein.
6. Wing JM. Computational thinking. Commun ACM. 2006;49(3):33–5. https://doi.org/10.1145/1118178.1118215.
7. Manovich L. Data science and digital art history. Int J Digit Art Hist. 2015;1:1. https://doi.org/10.11588/dah.2015.1.21631.
8. Hey T, Tansley S, Tolle K, editors. The fourth paradigm: data-intensive scientific discovery. 2nd ed. Redmond, Wash: Microsoft Research; 2009.
9. This diagram is based on the visualization in the following publication, Gomez C, Guardia A, Mantari JL, Coronado AM, Reddy JN. A contemporary approach to the MSE paradigm powered by artificial intelligence from a review focused on polymer matrix composites. Mech Adv Mater Struct. 2022;29(21):3076–96. https://doi.org/10.1080/15376494.2021.1886379.

Chapter 2
Setting Up R

To effectively analyze data, the choice of means is crucial. In this book, we are using the programming language R as a strategic choice, since it offers distinct advantages over alternative means. Here are the reasons for this choice:

R, being an open-source language, accommodates a wide spectrum of data analysis tasks. Its extensive community contributes to a growing ecosystem of packages and tools across diverse domains, enhancing its adaptability for cultural data exploration. R is particularly strong in data visualization with libraries like `ggplot2` and `plotly`, providing users with the ability to generate visually compelling and informative graphs. The expansive package repository equips R with a rich set of tools covering a broad spectrum of data analysis and statistical tasks. R's heritage as a statistical language positions it as a preferred choice for data analysts.

In the context of data analysis tools, Excel is recognized for its user-friendly point-and-click interface. It is widely utilized and offers a relatively low learning curve for basic tasks. However, mastering advanced features in Excel can be time-consuming. Excel is primarily oriented toward general spreadsheet tasks, lacking the flexibility inherent in a programming language. Furthermore, it imposes limitations on the amount of data it can handle in memory.

Python, on the other hand, is a versatile general-purpose programming language with extensive libraries catering to data analysis. In comparison, R's specialization in statistical operations often leads to the creation of more efficient and concise code for such analytical tasks. R's syntax and functions are frequently deemed more intuitive, particularly when it comes to data wrangling activities. The R community is tailored specifically to statistical computing and data analysis, fostering the development of highly specialized packages.

H. Klinke, *Cultural Data Science*, Quantitative Methods in the Humanities and Social Sciences, https://doi.org/10.1007/978-3-031-88130-5_2

RStudio serves as the graphical interface for the R programming language, providing a user-friendly environment for data analysis, visualization, and coding. When installing RStudio, it is important to understand the relationship between R and RStudio.

R itself is the core programming language, represented as the command-line interface typically seen in the lower-left corner of the RStudio window. In this context, RStudio can be considered as an integrated development environment (IDE) that enhances the functionality and user experience of R. It serves as a comprehensive platform for writing, running, and visualizing code.

Therefore, the recommended installation sequence is to first install R and then RStudio. R establishes the foundation, and RStudio is subsequently installed to provide a more interactive and feature-rich interface.

To download R and RStudio for your operating system, follow these steps:

1. First got to https://cran.rstudio.com/.
2. Download R by clicking on the appropriate version for your operating system, then choose "base."
3. Install R by double-clicking on the downloaded file.
4. After installing R, navigate to the following link: https://posit.co/download/rstudio-desktop/
5. Install RStudio. Note: If you are using a Mac and encounter a prompt asking to install "Git Developer Tools," it is not necessarily required for RStudio and can be skipped.
6. Once both R and RStudio are installed, launch RStudio.
7. Press Ctrl+1 to ensure that all four windows are visible and you should arrive at the result shown in Fig. 2.1.

> ℹ On a Macintosh, press Shift+0 or in the menu > View > Panes > Show all Panes. If you like, you can now adjust the window sizes by moving the mouse arrow to the horizontal or vertical dividing line until it shows a double arrow and moving it up/down or right/left.

By following these steps, you will have a complete setup with R and RStudio, ready to explore and analyze data in a seamless environment.

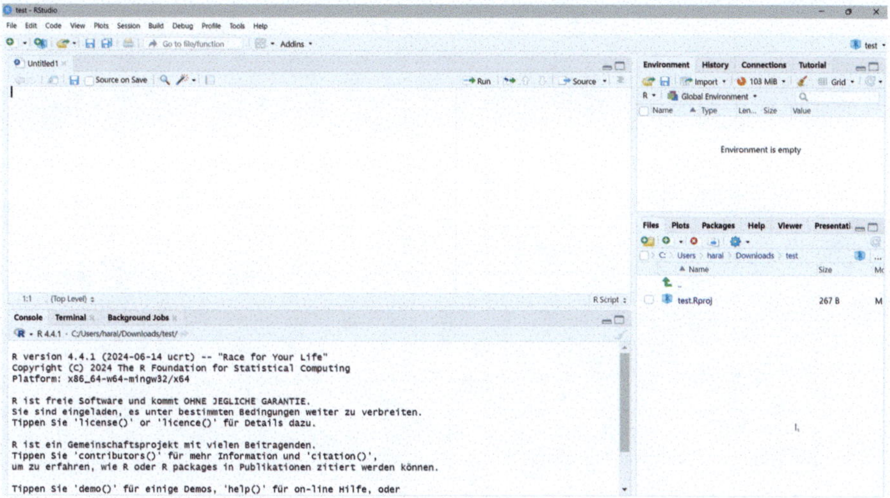

Fig. 2.1 The RStudio Interface

Chapter 3
First Steps

This chapter is dedicated to initiating your journey with R. We will explore the fundamental concepts of objects and operators, understand the efficiency boost provided by Code Completion, and unravel the basics of scripting, including executing scripts line by line or in their entirety. By the end, you will have a foundational understanding of R's essential components, setting the stage for more advanced data analysis and coding tasks.

Once you have RStudio installed, initiate a new project by clicking on the menu on File > New Project. In the subsequent window, choose New Directory > New Project (see Fig. 3.1). Enter the project name, which will also be the name of the subdirectory where all files will be stored. In this directory, you will find an initial `.Rproj` file. This project file serves as a configuration file, encapsulating project-specific settings, making it easier to manage and share your work. The creation of a new project helps maintain a well-organized and self-contained workspace for your analyses within RStudio.

Now, familiarize yourself with the windows, the look and feel, and the basic functions within RStudio.

In RStudio, there are four main windows. If the script window is not visible, press Ctrl+1 to reveal it. You can adjust the size of these windows by dragging the mouse cursor along the edges. Alternatively, use the symbols provided to minimize or maximize each window for a customized workspace.

Understanding the layout and functionality of these windows is essential for efficient navigation and utilization of RStudio. As you explore these features, you will gain a more comfortable and intuitive experience within the RStudio environment.

You can adjust the font size in RStudio by pressing Ctrl++ to increase or Ctrl+ – to decrease it. Alternatively, you can make these changes under Tools > Global Options > Appearance.

At the bottom left is the primary R console window, where you can interact with R by entering individual commands. To input commands in this window, click on it

H. Klinke, *Cultural Data Science*, Quantitative Methods in the Humanities and Social Sciences, https://doi.org/10.1007/978-3-031-88130-5_3

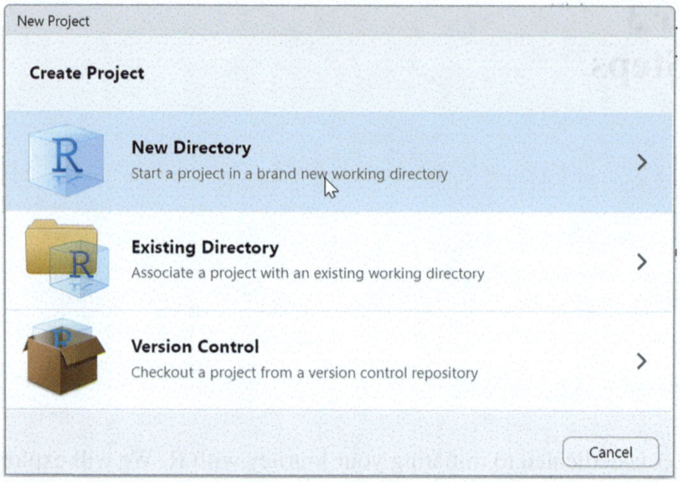

Fig. 3.1 Creating a New Project in RStudio

or press Ctrl+2. The blinking cursor indicates the insertion point where you can start typing.

Moving to the top left, we have the Script window. In this space, you can store a sequence of commands to be executed sequentially. To input commands in this window, simply click inside it or press Ctrl+1.

On the top right is the Environment window, which displays the data objects you create and want to access during your R session.

Finally, at the bottom right is the Files window. This area displays saved scripts and provides access to additional resources such as Help. For example, if you save a script, you will find it listed here. More details on these features will be explored later. Familiarizing yourself with these windows is essential for efficient navigation and productivity within RStudio.

3.1 The Console Window

Let us dive into the Console window for our initial exploration. Let us begin by entering a simple arithmetic operation: `1 + 1`, and press Enter. The result will be displayed below:

```
> 1+1
[1] 2
```

The number in brackets, `[1]`, indicates that this is the first result. The actual result, 2, follows.

This same approach can be applied to various operators:

+	Addition
-	Subtraction
*	Multiplication
/	Division
^ or **	Exponentiation
x %% y	Modulo (Remainder of Division)
x %/% y	Integer Division

While we are at it, let us introduce logical operators for comparing values. The result of a comparison is either "true" or "false."

<table>
<tr><td><</td><td>Less than</td></tr>
<tr><td><=</td><td>Less than or equal to</td></tr>
<tr><td>></td><td>Greater than</td></tr>
<tr><td>>=</td><td>Greater than or equal to</td></tr>
<tr><td>==</td><td>Exactly equal to</td></tr>
<tr><td>! =</td><td>Not equal to</td></tr>
<tr><td>!x</td><td>Not x</td></tr>
<tr><td>x|y</td><td>x or y</td></tr>
<tr><td>x&y</td><td>x and y</td></tr>
</table>

These basic operations lay the groundwork for more complex calculations and comparisons in R. As you explore and practice, you will become more adept at utilizing these operators in your data analysis tasks.

> ℹ The symbol | is called "Pipe" and is typically found on Windows keyboards on the bottom left, next to the Shift key in conjunction with the AltGr key. On a Mac press alt+7.

For instance, the following comparison yields "FALSE" because 1 is not greater than 2:

```
> 1 > 2
[1] FALSE
```

These comparison operators become crucial later when we aim to compare the contents of two data objects. This comparison result guides our decisions on which subsequent commands to execute based on the outcome of the comparison.

Now, let us create a data object to store data for later retrieval. This is a straightforward process with two possible notations: one uses "<–," and the other simply uses "=". While both are valid, "<–" is the more conventional approach in R, so we will use it in the following examples.

```
a <- 1
```

This creates an object named "a," and you will immediately see it in the Environment window on the top right. Switch from "List" to "Grid" and you will get this view (Fig. 3.2):

It is essential to pay attention to case sensitivity. For example,

```
A <- 1
```

would create another object named "A."

In the Environment window, you not only see the name of the object but also its content (Value), its type (numeric), and the memory it occupies. There are various data types which we will learn about later. For now, let us focus on one more aspect.

```
my_text <- "Hello"
```

This creates an object of type character (Fig. 3.3).

The distinction is clear: a sequence of letters is passed to the object in quotation marks.

```
my_text2 <- "1234"
```

Storing these digits as a character sequence means they are treated as text (Fig. 3.4). They cannot be directly used for arithmetic operations.

Now, objects can not only be created (written) but also displayed (read). To do this, simply enter the name of the object:

```
> my_text2
[1] "1234"
```

Now, let us explore how we can establish relationships between objects.

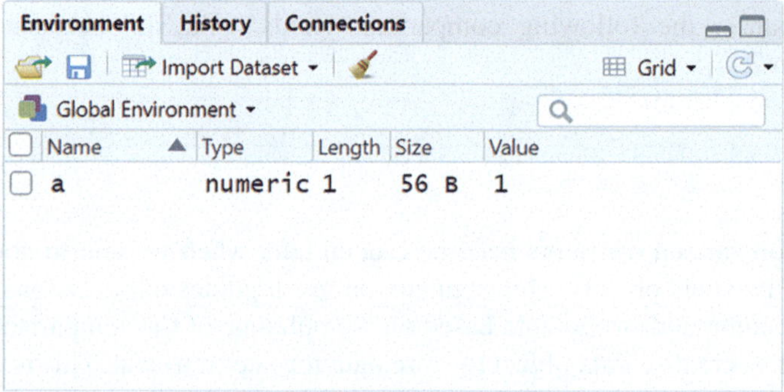

Fig. 3.2 An object in the Global Environment in RStudio

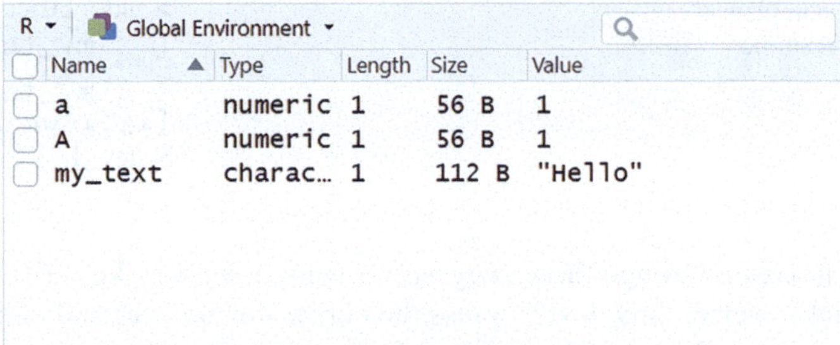

Fig. 3.3 Variables in the RStudio Global Environment

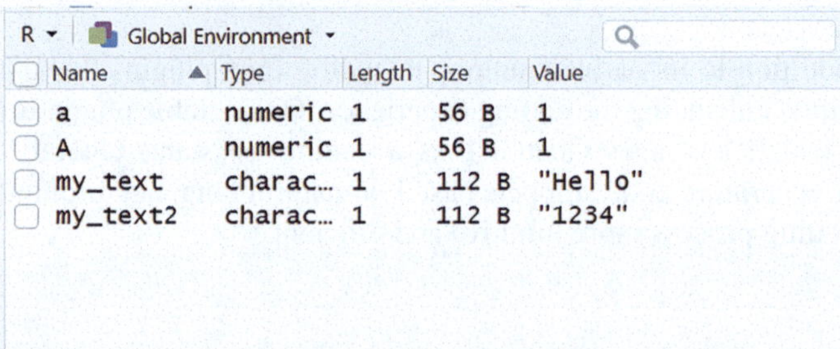

Fig. 3.4 Global Environment in RStudio showing numeric and character variables

```
> a+a
[1] 2
```

This corresponds to adding the content of "a" to itself, i.e., 1 + 1. Additionally, we can compare objects. As a reminder: One equals sign is an assignment, two equals signs mean a comparison.

```
> my_text == my_text2
[1] FALSE
```

This naturally results in "FALSE" as the outcome of the comparison because the contents are not the same.

```
> my_text == my_text
[1] TRUE
```

This logically evaluates to "TRUE" since we are comparing an object with itself.

If you have entered the commands, you may have already discovered the convenience of "Code Completion." When you start typing the initial characters of a known object, suggested objects may appear (Fig. 3.5).

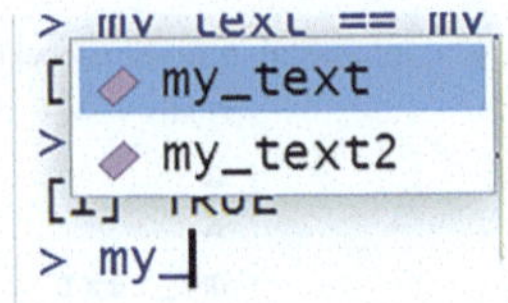

Fig. 3.5 Autocomplete suggestion in RStudio

You can navigate through these suggestions using the arrow keys (if, for example, you want to select "my_text2," press the cursor down once) and complete the text by pressing the Tab key. This feature allows for rapid code writing, especially when dealing with a potentially considerable number of objects, minimizing the risk of typos.

> In addition to its native features, RStudio offers plugins that extend its functionality, enhancing the coding experience. One notable plugin is GitHub Copilot, which leverages ChatGPT as a coding assistant. GitHub Copilot assists in generating code snippets based on natural language prompts, making the coding process more intuitive and efficient.

> For more detailed information on how to integrate and use GitHub Copilot in RStudio, you can refer to the documentation available here: https://docs.posit.co/ide/user/ide/guide/tools/copilot.html.

Moreover, "Code Completion" extends its functionality to functions. Functions act as mini programs, offering additional capabilities and enabling more complex operations with just a single line of code.

> We will explore how to extend R with additional functions or even create custom functions later. This involves delving into the process of expanding R's functionality by incorporating external packages or crafting your own functions tailored to specific needs.

For instance, when you type "ncha," you will be presented with the completion suggestion for the function "nchar." A yellow window also appears, providing a hint about the function's usage (Fig. 3.6). For now, we are only interested in the explanation that the function returns the number of characters in a character string (or string). This is achieved by passing the relevant string as an argument to the function. The name of the function is followed by parentheses, which then contain the argument, also called parameter. If you press the Tab key, the function name will be completed, and two parentheses will be automatically provided for you.

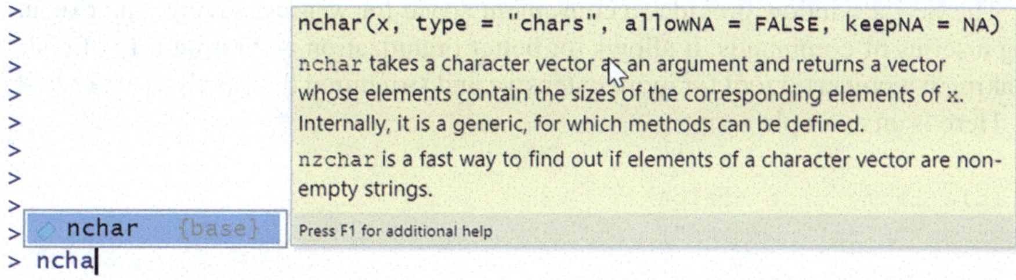

Fig. 3.6 Tooltip help in RStudio

This feature not only accelerates the coding process by suggesting relevant completions but also offers helpful insights into the functionality of the functions, making it easier for users to understand and utilize them effectively.

Now, let us use the function within our object.

```
> nchar(my_text)
[1] 5
```

The result indicates that the text consists of five characters.

Of course, we can also assign new values to objects.

```
> a = a + 1
```

Mathematically, this expression might seem odd. However, since the equal sign denotes an assignment, the value of "a" (which is 1) is incremented by 1 and written back into the object "a."

As described above, this is the reason the preferred notation in R for assignments is:

```
> a <- a + 1
```

This notation avoids confusion with the mathematical equality sign and clearly indicates that an assignment is being performed.

Now, the object a has the value of 2. This change is reflected in the Environment window on the top right.

3.2 The Script Window

Now, we may not want to input all commands directly into the Console window. Perhaps we would like to save them for future reference, execute multiple commands sequentially, or rerun a set of commands. This is where the Script window comes into play. A sequence of commands executed in order ultimately constitutes a script, and the Script window is designed to facilitate this.

The Script window provides a convenient space for writing, saving, and executing a series of commands. It allows for better organization and reusability of code, making it a powerful tool for more extensive and structured data analysis tasks in R.

Here is an example script:

```
# This is my first script
message("First command")
message("Second command :)")
name <- "My Name"
message(name)
message(nchar(name))
```

Type this code into the Script window on the top left.

To execute the code step by step, click on "Run" in the top right, or press `Ctrl+Enter`. (on a Mac: `Command+Enter`). You will see that this script demonstrates the basic of creating comments (using hashtags in the beginning of a line), executing commands, and utilizing variables in R Scripts.

To execute multiple lines or the entire code, highlight the respective lines or select all with `Ctrl+a`, then click "Run" or press `Ctrl+Enter`.

> ℹ It is essential to provide thorough comments in your code so that both yourself and others can understand the purpose and functionality of each section. To add comments, simply place a # symbol at the beginning of a line, and that line will not be executed.

> Comments serve as a form of documentation, explaining the logic or intention behind specific code segments. They enhance code readability and make it easier for anyone reviewing the script to grasp the structure and purpose of each part.

Save your script file regularly. You can do this by either clicking on the floppy disk icon or simply by using `Ctrl+S` (on a Mac: `Command+S`). Regular saving ensures that your work is preserved, preventing any potential loss of progress or changes made to your script.

Chapter 4
First Data

In this chapter, we are introducing a new data type called Data Frame, which is essentially a table. We will learn how to import an existing data set, apply basic analysis functions, and create meaningful data visualization.

To work with an initial set of collection data, let us download open data from the Museum of Modern Art.

"Open data" refers to data that is freely available for anyone to use, reuse, and redistribute without any restrictions on access or usage. The idea behind open data is to promote transparency, collaboration, and innovation by making datasets accessible to the public.

The Museum of Modern Art (MoMA) offers their collection data as open data on GitHub to make information about their artworks available to a wider audience. By providing open access to their collection data, MoMA encourages researchers, developers, and the general public to explore and analyze the information, fostering creativity and enabling a broader understanding of the artworks they have in their collection.

To download the collection data from MoMA's GitHub repository, you can follow these steps:

1. Visit the GitHub page: Go to the following URL in your web browser: https://github.com/MuseumofModernArt/collection/blob/master/Artworks.csv
2. Download the CSV file: On the GitHub page, you will see the message "(Sorry about that, but we can't show files that are this big right now.)". To download it, right-click on the "Raw" button (Fig. 4.1). Depending on your browser, this button might be labeled as "Save as…" or "Download."
3. Save the file: Once you click the button, the CSV file will be downloaded to your computer. Choose your RStudio project folder and save the file there, and you are done.

H. Klinke, *Cultural Data Science*, Quantitative Methods in the Humanities and Social Sciences, https://doi.org/10.1007/978-3-031-88130-5_4

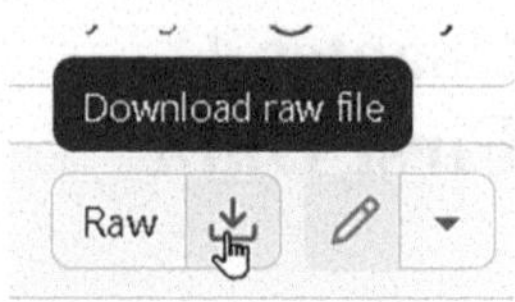

Fig. 4.1 GitHub interface showing the option to download a raw file

Now you have the Artworks.csv file from MoMA's collection available for exploration. This dataset will serve as our starting point for exploring and analyzing real-world data within the R environment.

The file you have downloaded is a CSV file, which stands for Comma-Separated Values. This means that individual values are stored in a plain text file, separated by commas. It is commonly used for storing tabular data, where each row represents a record, and the columns are separated by commas. CSV files are a simple and widely supported format for exchanging data between different software applications, databases, and systems.

In a CSV file:

- Each line typically represents a row of data.
- The values within a row are separated by commas (or other delimiters, such as semicolons or tabs).
- The first row often contains column headers, providing names for each column.
- Commas are used to delimit columns, but other delimiters may be used depending on the file.

For example, a CSV file might look like this:

```
Name, Year, Title
Albrecht Durer, 1500, Self-Portrait
Pablo Picasso, 1937, Guernica
Gerhard Richter, 1988, Betty
```

When importing the downloaded file into R, it needs to interpret this text with the correct character encoding, so it is recommended to choose UTF-8. To do this, go to the top right corner of the Environment window and click on "Import Dataset > From Text (base)." Select the downloaded file, select "UTF-8" under Encoding (Fig. 4.2). Then click on Import.

For a moment, you will see a red dot in the console window, indicating that R is busy. Afterward, a new object named "Artworks" will appear in the object window. You can observe that it contains over 130,000 rows representing artworks, described by twenty-nine properties (columns) (numbers may vary since this dataset is updated monthly).

Alternatively, if you want to ensure that the settings are correct, you can also directly use the following command for import:

Fig. 4.2 CSV import settings

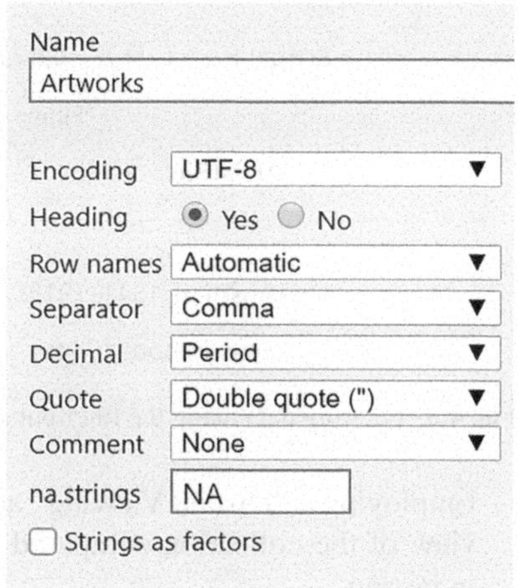


```
Artworks <- read.csv("Artworks.csv", encoding="UTF-8")
```

If the table is not displayed, click on the tabular symbol behind "Artworks" in the Environment (set to Grid) or enter the command `View(Artworks)` (with a capital 'V').

You can now:

- *Sort* the columns using the small triangles at the top edge.
- *Filter* the table using the button and entering a term in the column header field.
- *Search* across all columns using the search field at the top right.

First click the filter button so that the search line opens under the column names (Fig. 4.3). Now entering "Picasso" under "Artist," for example, will display all works to which Picasso has contributed. Under the table, you will see that there are 1322 objects related to Picasso.

> ⓘ You will probably get a different number of objects in the result. This is because the dataset is updated every month and new works are added. When you download it, it is already somewhat larger than it was at the time I drafted the book.

If you then enter "Avignon" under "Title," you will notice that not only the painting "Les Demoiselles d'Avignon" from 1907 is in the collection, acquired in 1939 (column: DateAcquired), but also three preliminary studies that came to MoMA later.

In this way, you will likely observe:

- There are empty values.
- There are works with multiple contributors.
- The "Date" column is a text field.

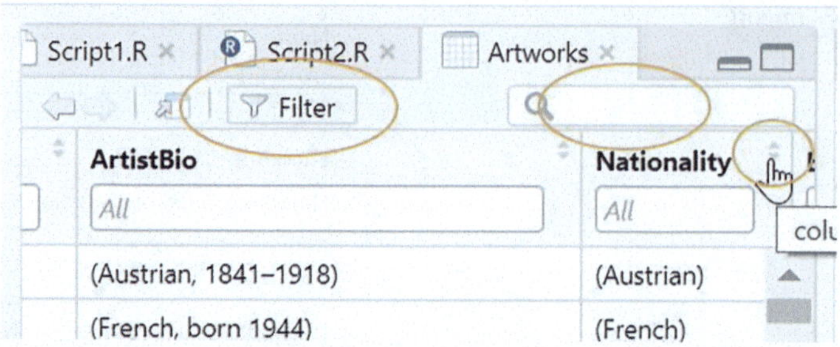

Fig. 4.3 Filtering data using the filter button, search bar, and column filters

– Employing a "Distant Viewing" approach provides a significantly different overview of the collection compared to what one would obtain by visiting MoMA in person.

You might want to ask the following questions:

1. What inquiries can we pose to such a data collection?
2. What responses can we anticipate from such a data collection?
3. What responses cannot be expected from it?
4. How will our approach to these data alter our questioning?

Keeping in mind that the Museum of Modern Art was founded in 1929, strongly associated with the "Modern" era, one might even assert that it was established as a museum for contemporary art at the time. Additional questions to consider include: How was the collection composed in its initial years? How has it evolved over the decades? How has the concept of "Modernism" changed during this period? How has MoMA contributed to shaping this concept through its collection policies? Which of these questions can be addressed using Data Science methods, and which cannot?

It is also important to acknowledge that these data were not initially intended for public use but rather for internal inventory purposes. Certain structures in the dataset can be explained by this origin, and it is reasonable to assume that not all internal objects have been publicly disclosed (considering issues related to restitution).

> i What methodological and substantive questions arise for you? Write down these questions and revisit them later when you have gained more experience with data and methodologies. This iterative process will likely lead to deeper insights and refinement of your questions as you become more acquainted with the intricacies of the dataset and analytical techniques.

Ultimately, you will have noticed that the search, sorting, and filtering capabilities provided by RStudio are quite limited. What we truly desire is programmatic access.

4.1 First Functions for Data Analysis

To start with, you can simply call the complete imported object (it is best to do this in the Script window, then you can execute the following commands one by one, repeatedly and if necessary modified):

```
Artworks
```

However, with 130,000 rows, displaying the entire dataset is impractical. RStudio shortens the display, offering little insight. Therefore, we need functions to gain a better overview.

Try the following to display the first few rows:

```
head(Artworks)
```

Similarly, to see the last rows:

```
tail(Artworks)
```

To find out how many rows (i.e., artworks) are in the table:

```
nrow(Artworks)
```

Similarly, you can check the number of columns:

```
ncol(Artworks)
```

Utilizing the "message" command, you can present a text to the user with the result, passing the elements as parameters separated by commas:

```
message("The dataset has ", nrow(Artworks), " artworks.")
```

If you want to see the names of all columns in the "Artworks" dataset in R, you can use the following command:

```
names(Artworks)
```

This command will display a list of column names in the console, showing you the variables or attributes present in the dataset.

If you wish to display only the "Artists" column, you write the column name separated by a dollar sign after the object:

```
Artworks$Artist
```

Of course, you do not want to display all, just the first few rows, to get a glimpse of what is stored in this column:

```
head(Artworks$CreditLine)
```

Now, if you are interested in not just the first few rows but a targeted sample of rows 2–5, you can use:

```
Artworks$Artist[2:5]
```

Note the colon here. Do not confuse this with the comma in square brackets, which fetches a value in a specific row and column:

```
Artworks[1,1]
```

These functions allow you to interactively explore your dataset in a more manageable way.

The arguably most helpful function for initially exploring an unfamiliar dataset is:

```
summary(Artworks)
```

This function provides insights into all the columns. Most of them have been imported as a string of characters, indicating there might not be much useful information there. However, those columns that exclusively contain numeric values have been imported as numerical types (pay attention to the dimensional details) (Fig. 4.4).

Isn't it fascinating? This function initially shows you the smallest value in each column (Min). For example, in the "Weight" column, the minimum value is 0.09 kg. On the other end, the heaviest object (Max) weighs an astonishing 185,067.59 kg (how was that weighed?).

> ⓘ As mentioned above, you probably have a more recent data set, so your values may differ from those shown here.

```
 Weight..kg.
Min.   :     0.09
1st Qu.:     5.67
Median :    19.96
Mean   :  1287.94
3rd Qu.:    80.29
Max.   :185067.59
NA's   :137892
```

Fig. 4.4 Summary statistics, displaying distribution metrics and missing values

It is crucial to note the presence of many NAs in the dataset. For instance, in the "Weight" column, there are 137,892 missing values. Out of the total 138,181 objects, only 289 have recorded weight information! This underscores the importance of handling missing data appropriately in subsequent analyses.

"NA" stands for "not applicable" or "not available," indicating an empty or missing value. The presence of NAs in a dataset can stem from several reasons, including being empty from the start, being left empty during the import or conversion processes. It is crucial to consistently be mindful of the diverse types of missing data, as we often can only make meaningful statements based on existing information.

Handling missing data is a critical aspect of data analysis. There are several strategies for dealing with NAs, such as removing rows or columns with missing values, imputing missing values based on statistical methods or domain knowledge, or using advanced techniques like machine learning algorithms that can handle missing data.

It is important to choose an appropriate method based on the specific context and the impact of missing data on the analysis. The decision should consider the overall goals of the analysis, the amount of missing data, and the potential biases that might be introduced by different handling strategies. Properly addressing NAs ensures that our analyses are robust and that we draw accurate conclusions from the available data.

Additionally, the summary function provides values for the first quartile (Q1) and the third quartile (Q3) when the data is sorted by magnitude. A quartile represents specific points in a data set that divide the data into four equal parts when sorted by magnitude. Moreover, it presents two measures of central tendency, namely the mean and the median. The reason for having two different measures is rooted in the distinct methods used for their computation.

The mean, or average, is calculated by adding up all the values and dividing by the total count of values. In this context, it can be influenced significantly by extreme values, as observed with the weight of the heaviest object, which pulls the mean to a relatively high value.

On the other hand, the median is determined by sorting all values and selecting the middle value. In a sorted list, the median marks the point where half the values are below and half are above. This provides a robust measure of central tendency,

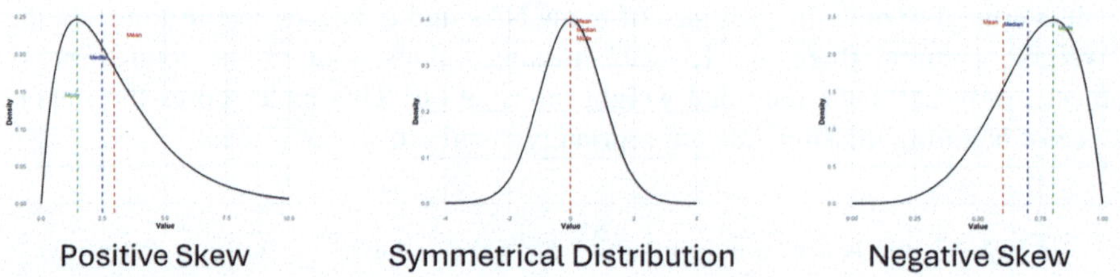

Fig. 4.5 Comparison of mean, median, and mode in different distributions

especially when the data set contains outliers or extreme values that might disproportionately impact the mean.

The inclusion of both terms "mean" and "median" in the summary offers a more comprehensive understanding of the distribution of values, enabling analysts to assess the central tendency from different perspectives and gain insights into the overall characteristics of the dataset.

Let us revisit these measures of central tendency (Fig. 4.5):

Mode represents the most frequently occurring value in a dataset. Identifying the mode can provide insights into the predominant characteristics of the data. However, in cases where no value repeats, the dataset is considered to have no mode.

In datasets following a normal distribution (common in measurement data), all measures of central tendency, particularly the mean, are equal. Normal distribution implies that the values are symmetrically distributed around the mean, creating a bell-shaped curve. Conversely, when a dataset deviates from normality, the mean may be influenced by outliers or skewed distributions, resulting in differences between the mean and other measures like the median.

In the context of our dataset, we suspect a concentration of values in the twentieth century (assuming the X-axis represents the year of creation), which could lead to different mean and median values. This observation underscores the importance of considering various measures of central tendency to capture the nuances of the dataset's distribution and gain a more accurate representation of its characteristics, especially when dealing with non-normally distributed data or datasets with unique patterns.

> Try it. What is the mean, median, and mode of the creation and acquisition dates?

Now let us pose some specific questions on the dataset:

What is the heaviest object in the collection?

```
min(Artworks$Weight..kg., na.rm = TRUE)
```

As simple as that is the lightest. The second parameter excludes the NAs.

And how heavy is the heaviest?

```
max(Artworks$Weight..kg., na.rm = TRUE)
```

But what object is it? We can find its index (essentially the row number) using:

```
which.max(Artworks$Weight..kg.)
```

The index is 75,020. Now, we can query our data object with this index:

```
Artworks$Artist[75020]
```

So, it is an object by Richard Serra. What is the title?

```
Artworks$ Title[75020]
```

It is called "Switch." And can we see it?

```
Artworks$ImageURL[75020]
```

This provides a URL to the image. Remove the quotation marks, copy it into the browser, and you will get the picture directly from the MoMA server: https://www.moma.org/media/W1siZiIsIjIxMDUzNiJdLFsicCIsImNvbnZ lcnQiLCItcmVzaXplIDEwMjR4MTAyNFx1MDAzZSJdXQ.jpg?sha=6416dcb5026f6e97

Isn't that great?

Could we express this differently? Are there not always multiple paths to the goal?

Yes, and here we can make it even more elegant and flexible:

```
Artworks$Title[ which.max(Artworks$Weight..kg.) ]
```

If you prefer, you can first write the index number into a variable and then pass it as a parameter:

```
index_heaviestwork <- which.max(Artworks$Weight..kg.)
Artworks$Title[index_heaviestwork]
```

Great! We are already becoming little data scientists.:).

Now, let us find the oldest object.

```
max(Artworks$Date)
```

What is happening here??

First, look at the results of these commands:

```
Artworks$Date[1]
Artworks$Date[10]
mode(Artworks$Date)
as.numeric(Artworks$Date[1])
as.numeric(Artworks$Date[10])
```

In the last line you will receive the warning "NAs introduced by coercion." It means that the conversion of the Date value in the specified row to a numeric format failed, resulting in an "NA" because the cell contains non-numeric data. We have noticed with Summary that the Date column is not numeric. So, max results in the end of an alphabetical sorting. That does not give us a meaningful result. What should we do now?

> Note that the contents of this column appear in quite different patterns. I have compiled an overview for you here, showing the TOP 20 patterns (out of a total of 1367) in the DATE field in the MoMA collection, which consists of 4320 patterns in 472,669 objects. Notice the differences between short and long hyphens, spaces and "n.d.", "c." and "c." - where there should be none (Fig. 4.6).

	Pattern	Freq	Example 1	Example 2
1	9999	86536	1896	1987
2	9999-9999	10292	1977-1978	1978-1980
3	9999-99	6622	1976-77	1980-81
4	a.w9999	5159	c. 1917	c. 1935
5	-9999	4959	1896	1987
6		2400		NA
7	9999,waaaaaaaaaw9999	1471	1934, published 1935	1932, published 1935
8	9999–99,waaaaaaaaaw9999	851	1930–35, published 1935	1967–68, published 1968
9	9999–9999	835	1958–1964	1985–1988
10	9999–99	773	1915–17	1972–78
11	a.w9999–99	753	c. 1925–30	c. 1928–29
12	a.a.	668	n.d.	NA
13	a.w9999-99	620	c. 1989-91	c. 1915-17
14	(9999-99)	597	1976-77	1980-81
15	a.9999	579	c.1985	c.1976
16	Aaaaaaw9999	516	Before 1977	Before 1955
17	(a.w9999)	484	c. 1917	c. 1935
18	9999a	462	1920s	1950s
19	9999/9999	355	1996/2003	1915/1945
20	Aaaaaw9999	310	After 1933	After 1924

Fig. 4.6 Frequency analysis of date patterns

We simply create a new column (`cdate`) where all convertible values from Date are written as a numeric value:

```
Artworks$cdate <- as.numeric(Artworks$Date)
```

Now, "ncol(Artworks)" should display one additional column: creation date as numeric.

> ℹ Note: Here, a warning message is generated. A warning is not an error. Instead, we expect a warning that not all values can be converted and that NAs will be created. This conversion is incomplete, but for our current purposes, it is sufficient.

That the result is numeric can be verified using:

```
mode(Artworks$cdate)
```

You can also see the result in the table at the very end: a new column with integers or NAs.

Now, knowing that we skip NAs, we can find the oldest artwork by querying the smallest number in our new column:

```
min(Artworks$cdate, na.rm = TRUE)
```

The result is 1768. Interesting, that is not exactly modern. Now we wonder which object that is. You already know how:

```
Artworks[which.min(Artworks$cdate),]
```

The comma here separates the row index from the column selection, with the empty space after the comma indicating that all columns should be included in the result.

And you also know how to find the average year:

```
mean(Artworks$cdate, na.rm = TRUE)
median(Artworks$cdate, na.rm = TRUE)
```

It is fantastic that you have stayed with it this far. But wait, the best part is yet to come.

How about visualizing the distribution of artworks over the years of creation?

Well, this actually calls for a histogram. Wikipedia tells us:

"A histogram is a graphical representation of the distribution of a cardinal scaled feature."

Let us try it:

```
hist(Artworks$cdate)
```

Wonderful? Please note that "hist" always requires numerical values. But we have created that with "cdate" now.

If you say that is a bit too coarse, then try it with the following additional parameter:

```
hist(Artworks$cdate, breaks = 1000)
```

> The parameter "breaks" in the "hist" function determines the number of intervals (or "bins") into which the data range will be divided. It influences the granularity of the histogram, affecting how the data is distributed across the x-axis. You can adjust this parameter to refine the visualization based on the characteristics of your dataset.

> To explore more details about the "hist" function and its available parameters, you can refer to the help documentation in R. You can access this information by typing the following command in your R console: `?hist`.

We can clearly see in Fig. 4.7 that we do not have a normal distribution but rather the expected bias in the twentieth century.

If you want to learn more about "hist" and its parameters, mark the command and press F1. You will see that a help text appears in the lower right window. There you will find detailed explanations, especially toward the bottom, with some usage examples. And this applies to all commands. Try it!

> If you like, also create a histogram for the acquisition date as well. First, we will create a new column `adate` and then generate the histogram.

In conclusion, the skills you have acquired here extend beyond understanding the Museum of Modern Art dataset and can be applied to any datasets you generate or import in the future. Taking a summary of your data is indeed good practice after importing. Also, do not forget to save your script using Ctrl-S. When you close RStudio, your entire project, including the data objects, will be saved, and reopened when you launch it next. This approach ensures that you can seamlessly pick up your work in subsequent sessions.

Fig. 4.7 Histogram, showing frequency distribution over time

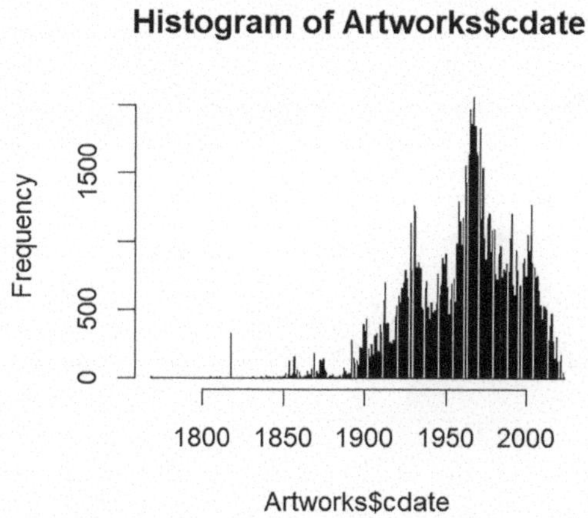

I am enough of the artist to draw freely upon my imagination. Imagination is more important than knowledge. Knowledge is limited. Imagination encircles the world.—Albert Einstein (1879–1955)[1]

[1] 1929 October 26, The Saturday Evening Post, What Life Means to Einstein: An Interview by George Sylvester Viereck, Start Page 17, Quote Page 117, Column 1, Saturday Evening Post Society, Indianapolis, Indiana.

Chapter 5
Exploring Data

In this chapter, we will explore additional methods to gain insights into data, focusing on diverse data visualization techniques and examining the contents of individual columns. We will also delve into the creation of subsets within the dataset.

To begin, we have just seen how to generate a histogram for numerical columns using the following command:

```
hist(Artworks$cdate, breaks = 1000)
```

This produces a histogram illustrating the distribution of converted data (excluding NAs) related to the creation dates (Fig. 5.1).

> ⓘ This represents the structure of the collection as of today. However, envision that the collection had a different fingerprint each year. For instance, a few years after the establishment of MoMA, the collection would have a distinct character compared to the 1980s. If we visualize this evolution as a histogram over time, you could implement an animation by creating subsets for each respective year, generating histograms for these subsets, exporting them, and compiling them into a clip. The process of creating subsets is explained at the end of this page, providing insights into how specific subsets were derived for each year.

Similarly, we could create a histogram for the acquisition date. However, you will notice that the first step leads to faulty data, as evident in the second step. The issue arises from the format of the "Date Acquired" column, which is in the format `YYYY-MM-DD`. Converting this directly to numbers results in errors. We have two

H. Klinke, *Cultural Data Science*, Quantitative Methods in the Humanities and Social Sciences, https://doi.org/10.1007/978-3-031-88130-5_5

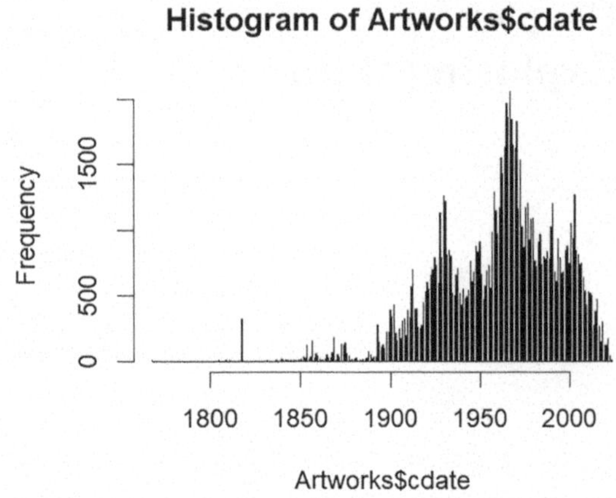

Fig. 5.1 Histogram of artwork creation dates from the MoMA dataset

potential solutions: extracting the first four digits (years) and converting them or converting the entire string into the Date format to preserve granularity.

When examining the "Date Acquired" column, you will notice the format is `YYYY-MM-DD`, with four digits for the year, two for the month, and two for the day, separated by hyphens. Converting this directly to numbers does not work as expected. Now, we have two options:

1. Extract the first four digits from the string and convert them, maintaining the year as a numerical value.
2. Convert the entire string into the Date format in a new column, preserving the existing granularity, i.e., acquiring the exact date.

The first option would look like this:

```
Artworks$adate <- as.numeric(substr(Artworks$DateAcquired, 1, 4))
```

Here, we use the "substring" function to extract characters from 1 to 4 and convert them into a numeric value. You can verify that the result is a number:

```
Artworks$adate[1] + 1
```

This demonstrates that the extracted value is numeric.

The second option can be implemented as follows:

```
Artworks$adate_date <- as.Date(Artworks$DateAcquired)
```

This converts the string into the Date format. Similarly, you can input the following and observe that the addition increases the date by one day (only displayed here, not written to the object):

```
Artworks$adate_date[1] + 1
```

Here, the result is a date value, and adding 1 increases the date by a day, illustrating the conversion to the Date format .

You can also do this for the year of birth and death. What did you notice? How would you address the following question: How old was the artist when the work was created?

You can follow a similar approach for birth and death years. What stands out is that you can calculate the age of the artist when the artwork was created by subtracting the birth year from the year of creation. Here is an example:

```
Artworks$Artist_Age <- Artworks$cdate -
as.numeric(substr(Artworks$BeginDate, 2, 5))
```

In this code, we create a new column "Artist_Age" in the "Artworks" data frame, representing the age of the artist when the artwork was created. The age is calculated by subtracting the artist's birth year (converted `BeginDate`) from the year the artwork was created (the calculated `cdate`).

> **?** If you type `summary(Artworks$Artist_Age)` you will see that it has produced some improbably high and negative data. Can you figure out why this is?

A histogram, as automatically generated, might lack visual appeal. To enhance it, you can add a title ("main") and clear axis labels ("xlab" for the x-axis and "ylab" for the y-axis). Here is an example in R:

```
# Create a histogram with title and axis labels
hist(Artworks$adate, breaks = 1000,
     main="Acquisitions by MoMA Each Year",
     xlab="Acquisition Date",
     ylab="Number of Artworks")
```

In this code, the "hist" function is used to create a histogram for the "adate" column in the "Artworks" data frame. The parameters "main," "xlab," and "ylab" are utilized to provide a title and axis labels for a more informative and visually appealing representation of the data. Adjusting these elements helps communicate the context and content of the histogram more effectively.

The result looks like this:

The histogram in Fig. 5.2 generated for the years based on the acquisition date, converted into the Date format, is created using:

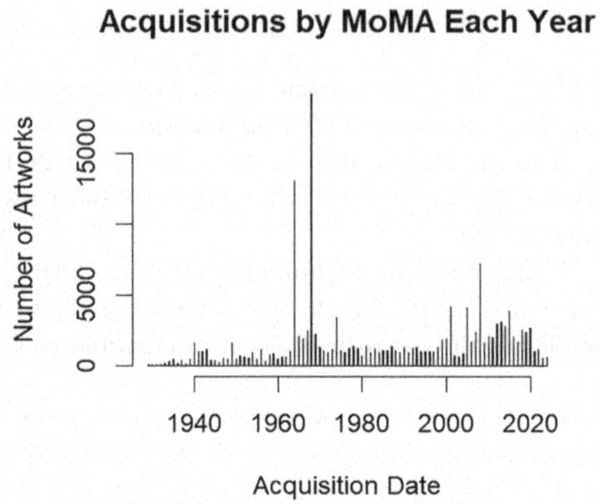

Fig. 5.2 Histogram of artworks acquired by MoMA each year

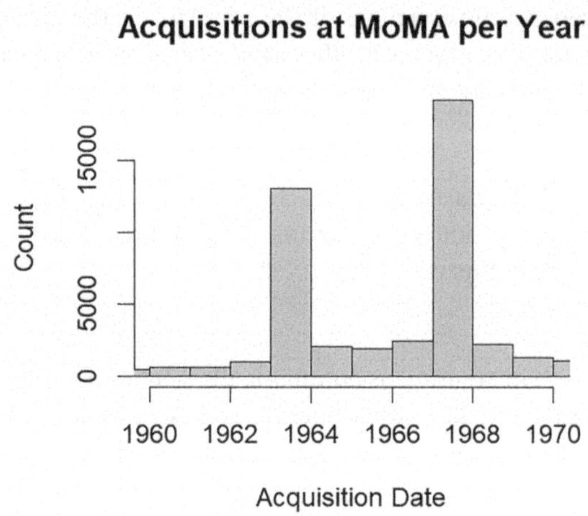

Fig. 5.3 Histogram of MoMA acquisitions per year

```
hist(Artworks$adate_date, breaks = "years")
```

? What are you "reading" from this diagram? What can such a representation reveal about a collection?

Interpreting this diagram involves examining the distribution of artworks over the years of acquisition. It provides insights into how the collection has evolved and the patterns of acquisitions over time.

Additionally, you can limit the axis to a specific time interval of interest, as demonstrated in Fig. 5.3:

```
hist(Artworks$adate, breaks = 100,
     main="Acquisitions at MoMA per Year",
     xlab="Acquisition Date",
     ylab="Count",
     xlim=range(1960:1970)
)
```

Another meaningful representation that visually depicts the values, similar to what we obtained with "summary," is the boxplot (or box-whisker-plot). The boxplot for the "cdate" (converted date) is created automatically without NAs (Fig. 5.4):

```
boxplot(Artworks$cdate, horizontal = TRUE)
```

This type of diagram illustrates the minimum and maximum values (excluding outliers) with vertical lines. The box represents the interquartile range, and the median is marked within the box. In this plot, the leftmost point indicates the artwork from 1768, which stands out as atypical for a modern art collection (Fig. 5.5).

ⓘ The term "interquartile range" refers to the range between the first quartile (Q1) and the third quartile (Q3) (i.e., the middle 50% of the data).

Fig. 5.4 Boxplot of artwork creation dates

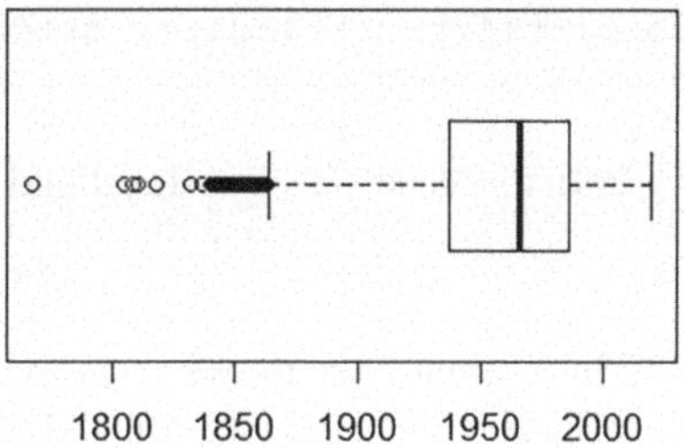

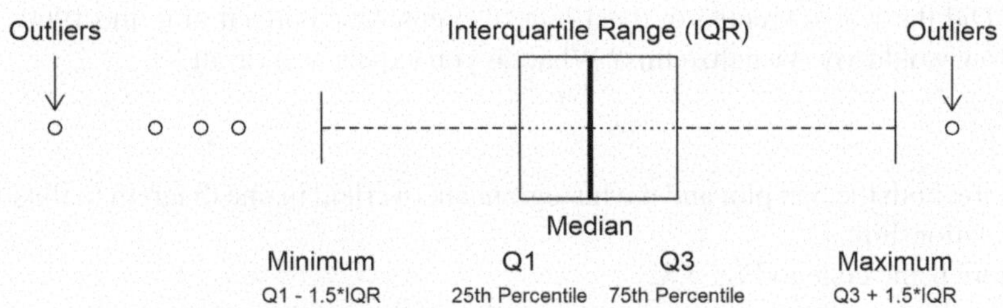

Fig. 5.5 Explanation of boxplot components

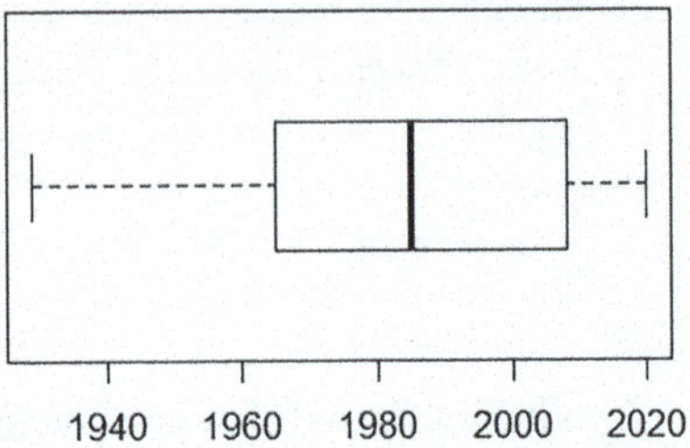

Fig. 5.6 Boxplot of artwork acquisition dates

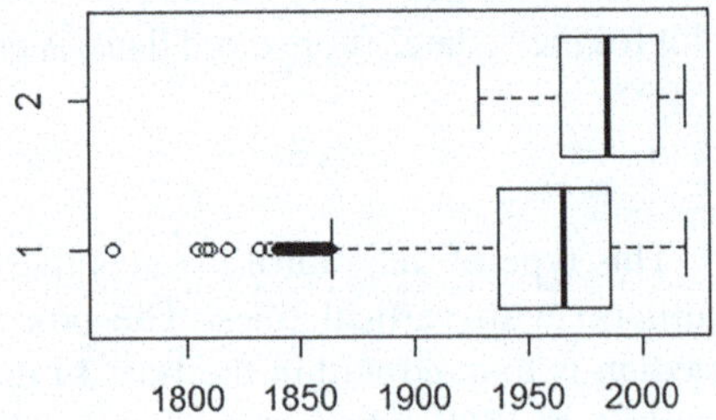

Fig. 5.7 Boxplot comparing artwork creation dates and acquisition dates

You can perform the same analysis on the acquisition date using (Fig. 5.6):

```
boxplot(Artworks$adate, horizontal = TRUE)
```

Or, for comparison purposes, you can display both the creation date and acquisition date in the same diagram (Fig. 5.7):

```
boxplot(Artworks$cdate, Artworks$adate, horizontal = T)
```

In this case, "T" is a shorthand for TRUE, and "F" would represent FALSE.

This comparison allows you to visually assess the distribution and potential relationships between the creation and acquisition dates of the artworks in the collection.

Did the works created in the 1960s also enter the collection in the 1960s? How would you visualize this? What do you expect as a result?

Here, both the box plot and the histogram are overlaid in one diagram to illustrate the relationship:

Here is the code to Fig. 5.8:

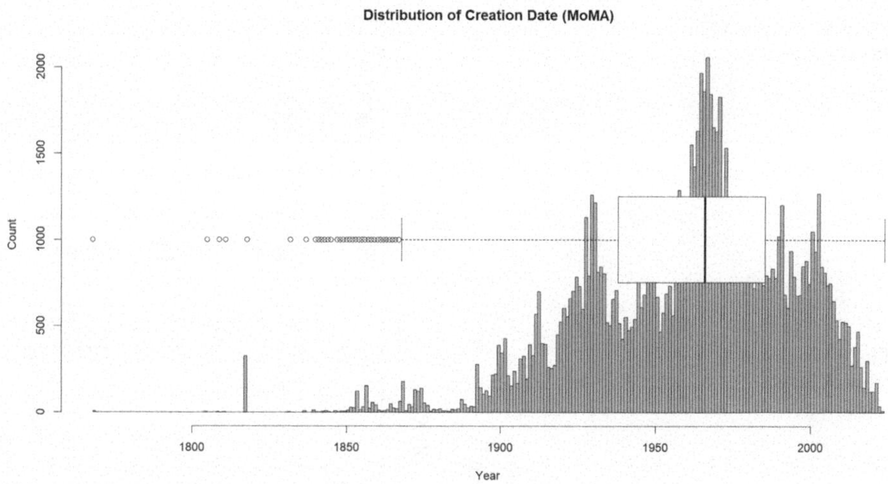

Fig. 5.8 Histogram and boxplot of artwork creation dates

```
hist(Artworks$cdate, breaks = 300,
     col = "gray",
     main = "Distribution of Creation Date (MoMA)",
     xlab = "Year",
     ylab = "Count")

# Overlay the boxplot horizontally
boxplot(Artworks$cdate,
        horizontal = TRUE,
        add = TRUE,
        at = 1000,    # Position below the histogram
        boxwex = 1000,   # Increase the width of the box
        col = "white",   # Set box color to solid white
        axes = FALSE)
```

? To what extent does this representation make the fundamental characteristics of a collection clear? How do you expect this representation to look like in other collections?

You can perform the same analysis with all other numerical data. For instance, here you can observe outliers in weight (Fig. 5.9):

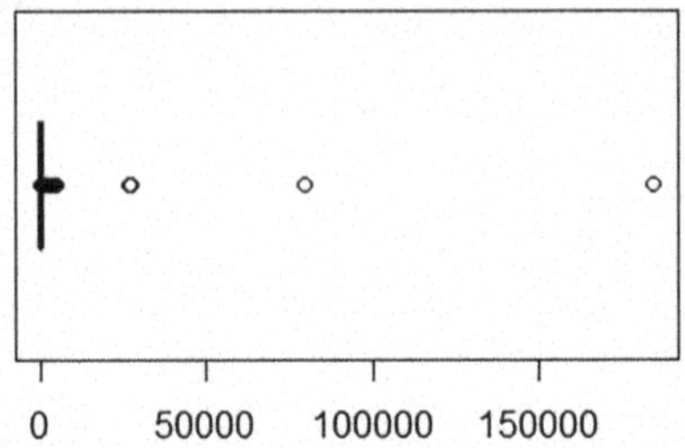

Fig. 5.9 Boxplot of artwork weights

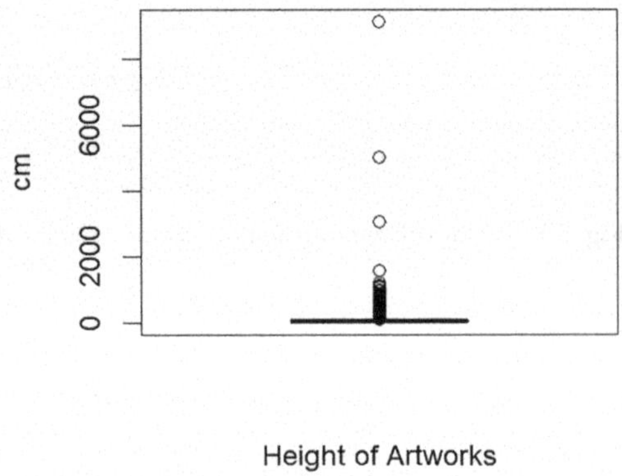

Fig. 5.10 Boxplot of artwork heights

```
boxplot(Artworks$Weight..kg., horizontal = TRUE)
```

And similarly, you can visualize which sizes are not very representative of the collection (Fig. 5.10):

5.1 Installing Additional Packages

Installing additional packages enhances the capabilities of our analysis. In RStudio, libraries, or packages, can be pre-installed or added as needed. For example, if we want to create a violin plot, we can install the "vioplot" package. To do this, navigate to the "Packages" tab in the lower right window, click "Install," search for "viop," and select "vioplot" (see Fig. 5.11).

Once installed, we can incorporate the functions from this package into our script by using the "library" command. For instance:

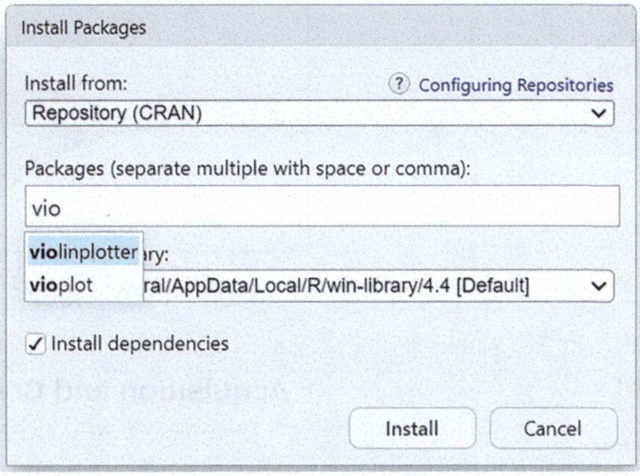

Fig. 5.11 Installing packages in RStudio

```
library(vioplot)
vioplot(Artworks$cdate, horizontal = TRUE)
axis(1)
```

A violin plot provides a smoothed histogram, displays the median and middle quantile. `Axis(1)` adds the default x-axis at the bottom (Fig. 5.12). We can use it to compare acquisition and creation dates (Fig. 5.13):

```
vioplot(Artworks$cdate, Artworks$adate,
        horizontal = TRUE,
        main="Acquisition and Creation Dates")
axis(1)
```

As a side note, the artwork from 1768 is Josiah Wedgwood's Coffee and Tea Service. To find this information programmatically, we can use:

```
min(Artworks$cdate, na.rm = TRUE)
Artworks[which.min(Artworks$cdate),]
```

The first line gives us the smallest year in the converted dates, and the second line retrieves the corresponding row in the "Artworks" dataset.

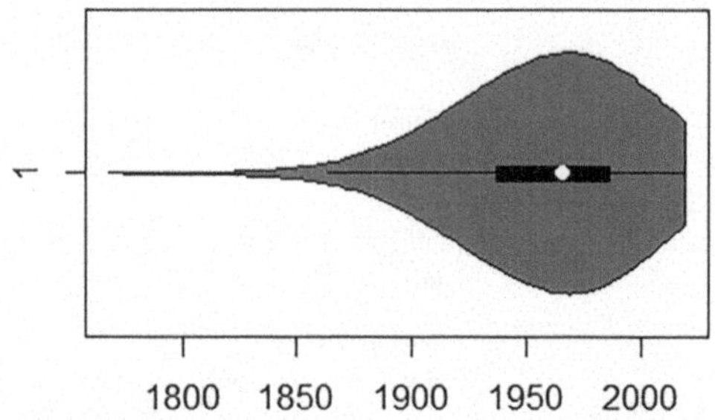

Fig. 5.12 Violin plot of artwork creation dates, visualizing distribution density over time

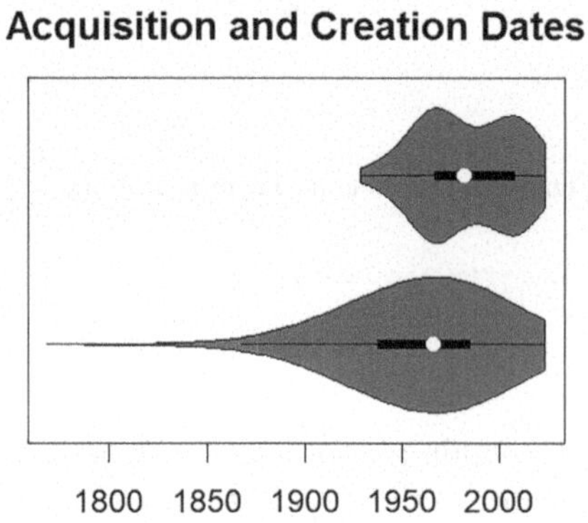

Fig. 5.13 Violin plot comparing artwork creation and acquisition dates

5.2 XY Diagram

Now, let us move beyond histograms and explore creating XY-plots to visualize the relationship between two existing values. For example, if a piece is 2 m wide and tall, we can mark a point at (2,2) on the XY-plot. This can be easily achieved with the following simple command:

```
plot(2, 2)
```

To extend this to all entries in our "Artworks" dataset, we can substitute the X and Y values with the "Width..cm." and "Height..cm." columns:

```
plot(Artworks$Width..cm., Artworks$Height..cm.)
```

To create Fig. 5.14, run the command. This may take a few seconds, as indicated by the red dot in the console window. The resulting plot will display the upper right corner of each artwork (Fig. 5.14), showing that most works are under 100 cm, with a few outliers exceeding 40 m in either length or width.

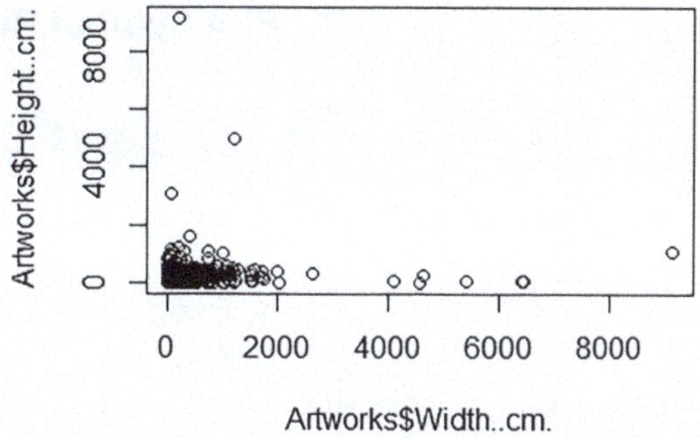

Fig. 5.14 Scatter plot of artwork dimensions, showing width versus height distribution

Note that this includes not only two-dimensional items but also three-dimensional works, where both the width and height are represented in the plot. This provides a comprehensive view of various dimensions in the dataset, capturing both planar and spatial aspects of the artworks.

Can we represent the third dimension in a plot? Perhaps, depicting the upper right corner with a circle may not make sense; a cross could be more meaningful. Fortunately, we can determine the symbol using the "pch" parameter (presumably standing for "point character"), as you can see in Fig. 5.15.

By limiting the axes, setting the transparency of the symbol to 25% (where "rgb" conveys the color value of red, green, blue, and alpha for transparency) and providing a title, the resulting plot is as follows:

```
plot(Artworks$Width..cm., Artworks$Height..cm.,
     pch = 3,
     xlim = range(0:100) ,
     ylim = range(0:100) ,
     col = rgb(0,0,0,.25),
     main = "Height and Width of Works in MoMA",
     xlab = "Width", ylab = "Height")
```

Executing the above code may take a while since all data points need to be processed and placed in the plot. The time required will depend on the performance of your device.

This visualization (Fig. 5.16) allows for a nuanced exploration of patterns that may not be apparent in a tabular format, revealing characteristics such as a diagonal of square works, variations above and below the standard, and irregularities in the collection.

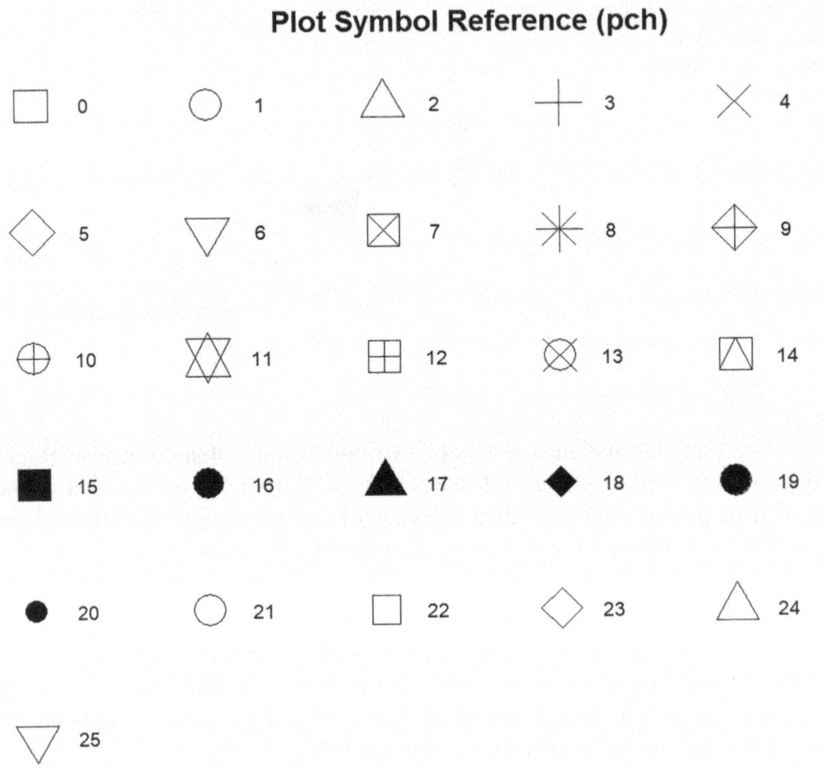

Fig. 5.15 Plot symbol reference chart for the pch parameter

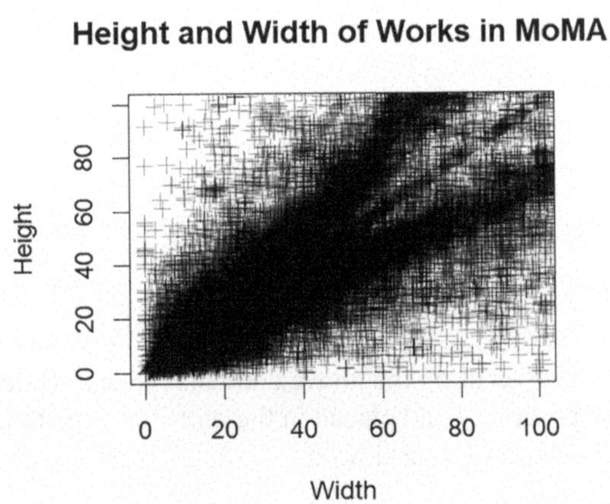

Fig. 5.16 Scatter plot of artwork height and width, showing the distribution and correlation of dimensions

The diagonal line indicates square works, while above and below it, you observe works in portrait and landscape orientations. Additionally, clusters of deviations from the standard become apparent, providing insights into the diverse dimensions and proportions present in the collection. This visual exploration enhances our understanding of the distribution and relationships within the dataset, allowing for a more intuitive grasp of the characteristics of the artworks.

? What does this representation tell us, and to what extent does it represent a profile of a collection? What other characteristics of the works can be represented in this way?

5.3 Exploring Properties

Now we have explored several columns. Which ones are still interesting? How about the "Gender" column? What different entries do you expect?

```
unique(Artworks$Gender)
```

This command provides a list of all unique values in the "Gender" column. There are many values; let us look at the first few:

```
head(unique(Artworks$Gender))
[1] "(Male)"                    "(Male) (Male)"           "(Male) (Female)"
      "(Male) (Male) (Male)"
[5] "() (Female) (Male)"    "(Female)"
```

Each involved person seems to have a field with entries separated by brackets. This structure might become important for us later.

Similarly, let us explore the MoMA departments:

```
unique(Artworks$Department)
```

Now, let us find out how many objects there are in each department:

```
table(Artworks$Department)
```

```
          Architecture & Design Architecture & Design -
Image Archive
                    20237                       30
             Drawings & Prints                     Film
                    77791                      874
             Fluxus Collection         Media and Performance
                     1938                     3006
           Painting & Sculpture                Photography
                     3981                    32991
```

This is interesting! Now, let us visualize this information:

```
plot(table(Artworks$Department))
```

What do you think? If you want a more esthetically pleasing version, try this (which results in Fig. 5.17).

```
plot(table(Artworks$Department),
     main="Number per Department",
     xlab = "Department", ylab = "Number",
     col = "blue")
```

Wonderful! Now, let us use the same approach for the "Medium" column:

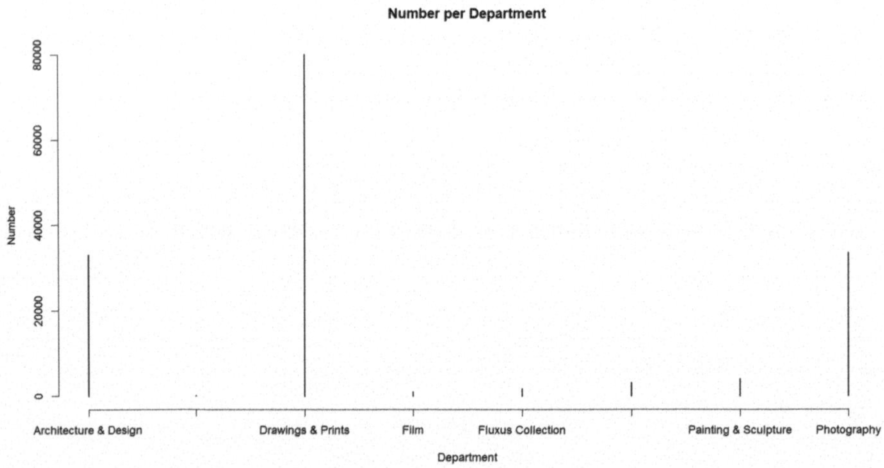

Fig. 5.17 Bar chart showing the number of artworks per department, highlighting major collection areas

```
unique(Artworks$Medium)
```

This column seems to have diverse content. Let us write the listing into a new table:

```
medium_table <- as.data.frame(table(Artworks$Medium))
View(medium_table)
```

Now, let us sort it by frequency:

```
medium_table <- medium_table[order(medium_table$Freq, decreasing
= TRUE),]
```

And display the most frequent entries:

```
print(head(medium_table), row.names=FALSE)
                 Var1  Freq
 Gelatin silver print 16638
                       9631
           Lithograph  7864
 Albumen silver print  4840
      Pencil on paper  1847
     Chromogenic print  1734
```

It appears that we have a considerable number of photographs and prints.

So, step by step, we are gaining a better understanding of the collection. Now, let us explore the "Classification" column:

```
table(Artworks$Classification)
```

Finally, let us look at the Nationality column:

```
# Write to a list and display the first 10
nat_list <- unique(Artworks$Nationality)
nat_list[1:10]
```

This gradual exploration is helping us form a more comprehensive picture of the collection.

> In R, a list is a versatile data structure that can hold elements of different types (e.g., numbers, strings, vectors, and even other lists) in a single object.

5.4 Subsets

So, now we have gathered a lot of information about the collection as a whole. However, perhaps we want to extract subsets from this overall collection to examine them individually.

For example, if I were interested only in Picasso, I could create a subset as follows:

```
MoMa_Picasso <- subset(Artworks, Artist == "Pablo Picasso")
```

The “subset” function takes the dataset from which a subset should be taken and a condition for this subset. Here, we define that the “Artist” column should be exactly “Pablo Picasso”. If this condition is true, the row is part of the subset; otherwise, it is not. If the condition is true, then yes; if false, then no.

> ℹ Note the double equals here. It is a comparison operator.

Now we have another, smaller dataset. Everything we have done with the complete dataset, you can now apply to this one: Summary, histogram, boxplot, XY-plot, table, and even create further subsets.

Try it:

```
summary(MoMa_Picasso)
head(MoMa_Picasso)
tail(MoMa_Picasso)
nrow(MoMa_Picasso)
View(MoMa_Picasso)
unique(MoMa_Picasso$Artist)
```

These commands allow you to explore and summarize the specific subset of artworks by Pablo Picasso.

5.5 Subsets Based on Lists

If you are interested in Picasso and Braque, you can use the %in% operator to specify a list of names:

```
MoMa_Picasso <- subset(Artworks, Artist %in% c("Pablo Picasso",
"Georges Braque"))
```

> ℹ `%in%` checks if elements of a vector match any of the values in the following vector.

And if we are looking for a list of people to be included in the subset, we can create that list first:

```
search_words <- c("Picasso", "Braque")
subset_artworks <- subset(Artworks,
                          grepl(paste(search_words, collapse="|"),
                                Artworks$Artist))
```

This results in 1499 objects.

> The paste() function is used to concatenate the elements of the search_terms vector into a single string.

If you want to create a subset based on a column from another table, you can use that column as a list for comparison.

> ? Look at the presentations of the teams from the National Gallery of Art Datathon.[1] Team 4 mentioned a Gender App in their presentation. What did these teams discover? Can we generate something similar with our MoMA dataset?

Note that MoMA provides not only collection data, but also exhibition data. You can find MoMAExhibitions1929to1989.csv and MoMADirectorsDepartmentHeads.csv under Exhibitions. Try downloading and importing these datasets. Let us know what you find in them.

On GitHub, you will also find a dataset from the Metropolitan Museum of Art. MetObjects.csv contains 470,000 works. So, be cautious when handling the entire dataset.

Look at the columns and their contents in the Met dataset. What criteria could be used to create meaningful subsets here? What might interest you in this museum?

[1] https://www.nga.gov/press/2019/datathon.html

Chapter 6
Compare Data

This chapter provides a comprehensive overview of programming structures in R, emphasizing loops, conditional statements, and subsets creation.

6.1 Programming Structures

Programming fundamentally involves repeating certain tasks through loops, making decisions based on conditions, and comparing different elements. These core structures—loops, decisions, and comparisons—are fundamental to all programming languages.

The advantage of programming lies in its efficiency, notably in saving time by automating repetitive tasks using loops. This capability allows for the concise and systematic execution of code, making it a powerful tool for handling complex operations.

6.2 Loops

Loops in programming follow a specific structure, often exemplified by the for loop:

```
for (variable in vector) {
}
```

In this construct, the variable, typically denoted as “i” for iteration, increments with each iteration (and is thus always numeric). The vector specifies the range or conditions for the loop to run; for instance, running ten times would be represented

H. Klinke, *Cultural Data Science*, Quantitative Methods in the Humanities and Social Sciences, https://doi.org/10.1007/978-3-031-88130-5_6

by “1:10.” The commands to be executed repeatedly are enclosed within the curly braces. For example:

```
for (i in 1:4) {
    message(i)
}
```

In this example, the loop iterates four times, and the command within the loop outputs the value of “i” during each iteration, resulting in:

```
1
2
3
4
```

This structure allows for the efficient repetition of specific actions, contributing to the concise and systematic execution of code.

6.3 Decisions

In programming, decision-making is a crucial aspect that allows the execution of code based on specified conditions. The basic structure for decision-making in many programming languages is expressed through the “if” statement:

```
if (condition) {
  # Code to be executed if the condition is true
} else {
  # Code to be executed if the condition is false
}
```

Here, the condition serves as the decision point. If the condition evaluates to true, the code within the curly braces after “if” is executed; otherwise, the code following “else” is executed. The “else” part is optional.

For example:

```
i <- 4
if (i == 3) {
  message("is three")
} else {
  message("is not three")
}
```

The result would be:

```
is not three.
```

This is because the value of "i" is 4, and thus, only the code within the "else" block is executed.

You can combine decision-making within loops. In the following loop example:

```
for (i in 1:4) {
  if (i == 3) {
    message("is three")
  } else {
    message("is not three")
  }
}
```

The result would be:

```
Is not three.
Is not three.
Is three.
Is not three.
```

This demonstrates that during each iteration of the loop, the value of "i" is checked, and the appropriate message is displayed.

For more complex scenarios involving comparisons against multiple values, nested if statements are an option, but a more concise approach is using the "switch" statement:

```
switch(object,
  case = action
)
```

Two application examples of the switch statement are provided below for further illustration.

6.4 Comparisons

Often, visual comparisons are essential in data analysis, such as comparing subsets of the overall dataset or two collections against each other. One way to achieve this is by creating side-by-side plots (Fig. 6.1):

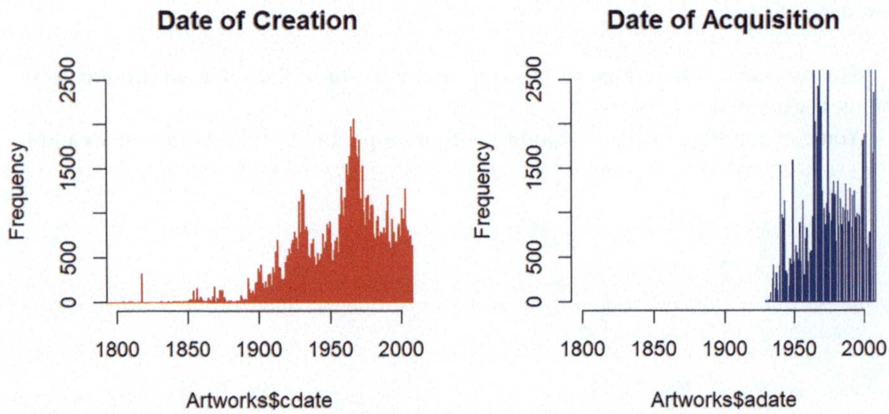

Fig. 6.1 Side-by-side histograms comparing artwork creation dates and acquisition dates, highlighting temporal trends

```
par(mfrow=c(1,2))  # One row, two columns
hist(Artworks$cdate,
     xlim=range(1800,2000), ylim=range(0:2500), breaks=400,
     col="red", border="red", main="Date of Creation")
hist(Artworks$adate,
     xlim=range(1800,2000), ylim=range(0:2500), breaks=400,
     col="blue", border="blue", main="Date of Acquisition")
```

The "par" function is used to define the layout, specifying the number of rows and columns. In this example, histograms of "cdate" (date of creation) and "adate" (date of acquisition) are juxtaposed. If the plots appear too close, adjustments to the margins can be made to avoid the "figure margins too large" error:

```
par(mar=c(1,1,1,1))
```

To reset the plot properties, the following command can be used:

```
dev.off()
```

For comparative purposes, it might also be beneficial to overlay two plots (Fig. 6.2). Ensure that the axes are similarly defined in both plots:

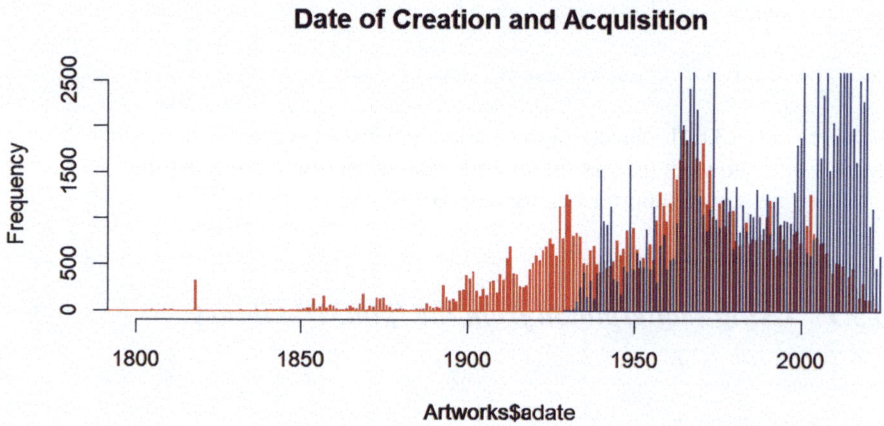

Fig. 6.2 Overlapping histogram, illustrating trends and time gaps between creation and acquisition

```
hist(Artworks$cdate,
     xlim=range(1800,2020), ylim=range(0:2500), breaks=400,
     col="red", border="red",
     main="Date of Creation and Acquisition")
par(new=T)
hist(Artworks$adate,
     xlim=range(1800,2020), ylim=range(0:2500), breaks=400,
     col=adjustcolor("blue", alpha.f=.5),
border=adjustcolor("blue", alpha.f=.5),
     main="Date of Creation and Acquisition")
```

Here, the colors of the bars and their borders are defined. To ensure that the second plot does not replace but overlays the first one, the "par" function is utilized. The colors of the bars in the second plot are set to be semi-transparent ("alpha.f=0.5"), allowing the bars from the first plot to remain visible.

6.5 Third Dimension

With an XY plot, we can visualize two series of values. However, what if we have three series of values, such as height, width, and depth?

First, let us examine how many non-empty values are in the depth column:

```
sum(!is.na(Artworks$Depth..cm.))
# or
length(na.omit(Artworks$Depth..cm.))
```

There are 13,841 objects, approximately 10% of the artworks, which have a depth value. Now, let us explore several ways to represent these depths.

There are several approaches we can take:

6.5.1 Using scatterplot3d from the 'plot3D' Library

```
library("plot3D")
scatter3D(Artworks$Width..cm.,
          Artworks$Depth..cm.,
          Artworks$Height..cm.,
          pch="+", col=rgb(0,0,0,.25),
          xlab="Width", ylab="Depth", zlab="Height",
          xlim=range(0:150), ylim=range(0:150),
zlim=range(0:150),
          bty="g")
```

> ⓘ If you install the plot3D library on the Mac, you also need to install xquartz, then restart RStudio and it will work.[1]

Here, we utilize the "scatter3D" function from the "plot3D" library to create a 3D scatter plot, representing width on the x-axis, depth on the y-axis, and height on the z-axis (Fig. 6.3). The points are marked with a cross symbol, and their color is semi-transparent.

> ⓘ Missing values are simply skipped, and no data point is displayed.

[1] https://www.xquartz.org/

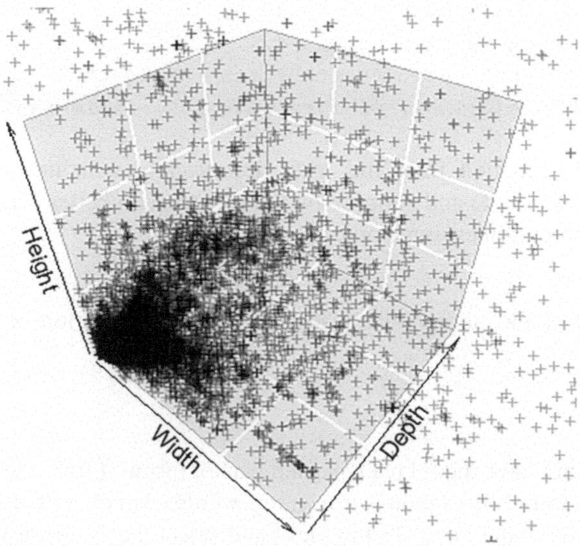

Fig. 6.3 3D scatter plot of artwork dimensions

6.5.2 Using scatterplot3d from the 'scatterplot3d' Library

```
library("scatterplot3d")

scatterplot3d(Artworks$Width..cm.,
              Artworks$Depth..cm.,
              Artworks$Height..cm.,
              xlab="Width", ylab="Depth", zlab="Height",
              xlim=range(0:150), ylim=range(0:150),
zlim=range(0:150),
              pch=16, color=adjustcolor("steelblue",
alpha.f=0.2))
```

Here, the "scatterplot3d" function is employed from the "scatterplot3d" library to generate a similar 3D scatter plot. The points are represented by solid circles, and their color is adjusted to be semi-transparent steel blue.

6.5.3 Using plot3d from the '`rgl`' Library

```
library(rgl)

plot3d(Artworks$Width..cm.,
       Artworks$Depth..cm.,
       Artworks$Height..cm.,
```

```
size=.7, col="black", alpha=.5,
main="MoMA (excluding NAs)",
xlab="Width", ylab="Depth", zlab="Height",
xlim=range(0:150), ylim=range(0:150), zlim=range(0:150))
```

The "plot3d" function from the "rgl" library creates an interactive 3D plot (Fig. 6.4). You can rotate and zoom in on the plot to get a better perspective of the distribution of three-dimensional objects.

> Also the plot3d library might need the initial installation of xquartz on the Mac.[2]

> If you do not have the rgl package and cannot install it from the repository, download it from https://cran.r-project.org/web/packages/rgl/For Mac users, choose the file under "macOS binaries" and select the r-release version. Do not unpack the file; instead, install it in RStudio using the "Install from Package Archive File" option. Choose the file and click "Install." Make sure you have xquartz installed.

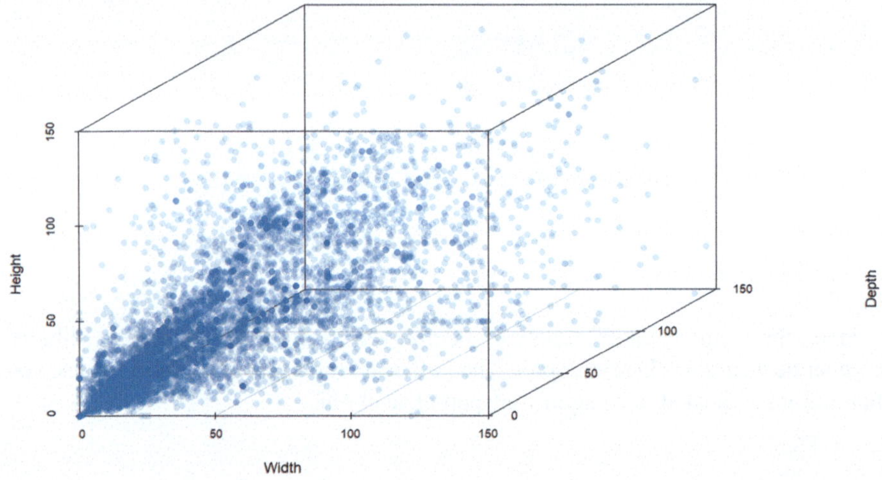

Fig. 6.4 3D scatter plot of artwork dimensions, illustrating size distribution and relationships

[2] https://www.xquartz.org/

The fascinating aspect of this chart (Fig. 6.5) is its interactivity. You can rotate and even zoom in on it!

When viewed from the front, it resembles a classic XY plot. However, you can also examine it from the side and top, gaining insights into the distribution of three-dimensional objects.

> Caution: The function does not seem to accept German umlauts in text values.

It would be even more appealing if the points were not just black but carried information about their department affiliation. Let us enhance the plot with meaningful information. First, let us review all the departments:

```
table(Artworks$Department)
```

Now, let us iterate through the table, creating a new column with color names based on the department. Here, the loop and decision structures come in handy:

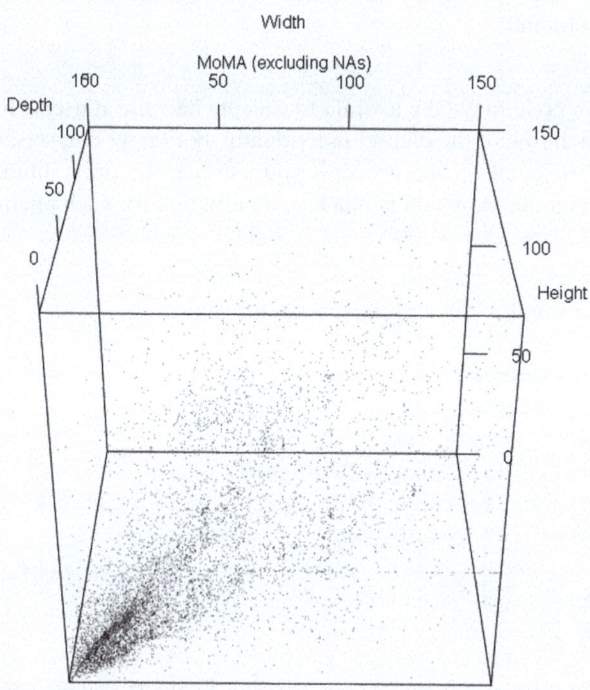

Fig. 6.5 3D scatter, excluding missing values, to show the distribution of physical sizes

```
for (i in 1:nrow(Artworks)) {
  switch(Artworks$Department[i],
         "Architecture & Design" = {Artworks$dep_col[i] = 
"blue"},
         "Drawings & Prints" = {Artworks$dep_col[i] = "gray"},
         "Painting & Sculpture" = {Artworks$dep_col[i] = "red"},
         "Photography" = {Artworks$dep_col[i] = "black"},
         "Media and Performance" = {Artworks$dep_col[i] = 
"yellow"},
         "Film" = {Artworks$dep_col[i] = "green"},
         "Architecture & Design - Image Archive" = {Artworks$dep_
col[i] = "blue"},
         "Fluxus Collection" = {Artworks$dep_col[i] = "orange"},
         {Artworks$dep_col[i] = "sandybrown"} # in case none of 
the above conditions are met (which should never happen in 
this case)
  )
}
```

This loop goes through a switch statement: if the department is "Architecture & Design," assign "blue" to the dep_col column, and so on.

The execution may take some time. This only needs to be executed once to create the _dep_col column.

> ℹ The above code may take a while to execute because it uses a loop, which processes each row of the dataset individually. For larger datasets, this can be very slow. To speed up the process, you can use the dplyr library and the case_when command, which is much more efficient for such operations.

Now, we can use this new column as a color value:

```
plot3d(Artworks$Width..cm.,
       Artworks$Depth..cm.,
       Artworks$Height..cm.,
       size=2.7, col=Artworks$dep_col,
       main="MoMA (without NAs) by Department",
       xlab="Width", ylab="Depth", zlab="Height",
       xlim = range(0:150), ylim = range(0:150), zlim = 
range(0:150),
       alpha=.9)
```

We note in Fig. 6.6 that many objects have no depth. Therefore, we would like to limit the number of objects to those with a depth greater than 1 cm. We can achieve this by either creating a new subset or defining this limit directly in the plot:

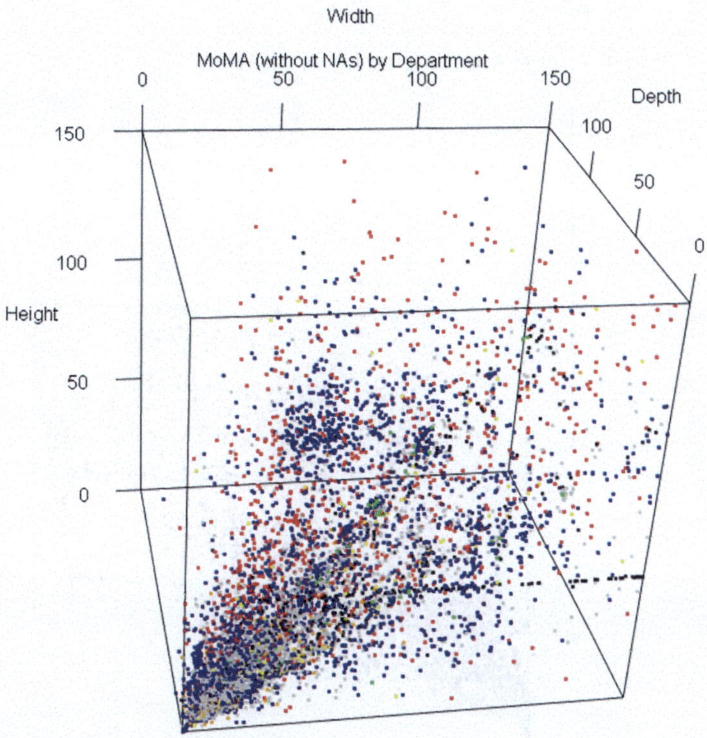

Fig. 6.6 3D scatter, colored by department, to show distribution by category

```
plot3d(Artworks$Width..cm.,
       ifelse(Artworks$Depth..cm. > 1, Artworks$Depth..cm., NA),
       Artworks$Height..cm.,
       size=2.7, col=Artworks$dep_col,
       main="MoMA (without NAs, >1cm) by Department",
       xlab="Width", ylab="Depth", zlab="Height",
       xlim = range(0:150), ylim = range(0:150), zlim =
range(0:150),
       alpha=.9)
```

In this case, a point is only plotted if the depth is greater than 1 cm; otherwise, it is skipped (Fig. 6.7). This does not visibly change the output, but speed up the rendering.

Now, it becomes clear that most points are blue, indicating the Architecture and Design department. Let us take a closer look by creating a subset that is limited to this category and only includes rows where the depth is greater than 0 (i.e., not NA):

```
MoMa_arch3d <- subset(Artworks, Department=="Architecture &
Design" & Depth..cm. > 0)
nrow(MoMa_arch3d)
```

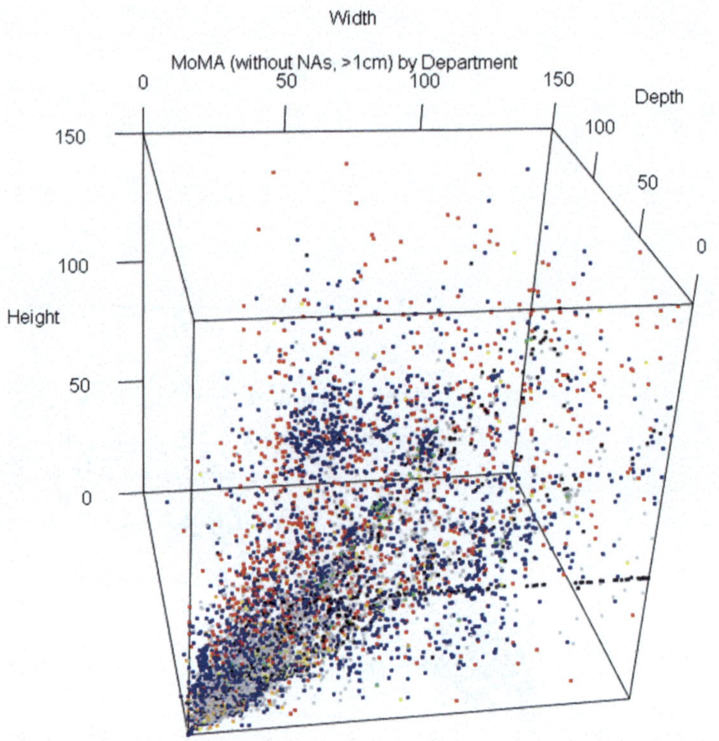

Fig. 6.7 3D scatter, excluding missing values and dimensions less than 1 cm

This subset contains 2416 objects.

Since they all have the same color now, let us use another attribute to define the color. We create another column for colors based on the classification:

```
table(MoMa_arch3d$Classification)


for (i in 1:nrow(MoMa_arch3d)) {
  switch(MoMa_arch3d$Classification[i],
         "Architecture" = {MoMa_arch3d$class_col[i] = "black"},
         "Design" = {MoMa_arch3d$class_col[i] = "blue"},
         "Frank Lloyd Wright Archive" = {MoMa_arch3d$class_col[i]
= "red"},
         "Mies van der Rohe Archive" = {MoMa_arch3d$class_col[i]
= "yellow"},
  )
}
```

Now, let us plot the Architecture and Design objects with color based on classification:

```
plot3d(MoMa_arch3d$Width..cm.,
       MoMa_arch3d$Depth..cm.,
       MoMa_arch3d$Height..cm.,
       size=4, col=MoMa_arch3d$class_col,
       xlab="Width", ylab="Depth", zlab="Height",
       main="Architecture & Design by Classification",
       alpha=.5)

text3d(MoMa_arch3d$Width..cm.,
       MoMa_arch3d$Depth..cm.,
       MoMa_arch3d$Height..cm.,
       texts = round(MoMa_arch3d$Depth..cm., digits = 0),
       cex=.6, col="black", alpha=.5)
```

This is what appears in a separate window:

Here, blue indicates Design (Fig. 6.8). “text3d” adds labels to the points. Here, it places rounded depth values as labels on each point in the 3D plot.

Now, let us enhance the visibility and limit the text to values deeper than 3 m (Fig. 6.9):

```
plot3d(MoMa_arch3d$Width..cm.,
       MoMa_arch3d$Depth..cm.,
       MoMa_arch3d$Height..cm.,
       size=4, col=MoMa_arch3d$class_col,
       xlab="Width", ylab="Depth", zlab="Height",
       main="Architecture & Design/Depth",
       alpha=.4)
text3d(MoMa_arch3d$Width..cm.,
       MoMa_arch3d$Depth..cm.,
       MoMa_arch3d$Height..cm.,
       texts = ifelse(MoMa_arch3d$Depth..cm. > 300,
                      paste0(round(MoMa_arch3d$Depth..cm., digits
= 0),"cm"),
                      ""),
       cex=.8, col="black", alpha=.9)
```

With `paste0`, we concatenate (without any separator between them) the rounded depth value with the string “cm,” creating a label that could make sense to use for Classification as well (Fig. 6.10).

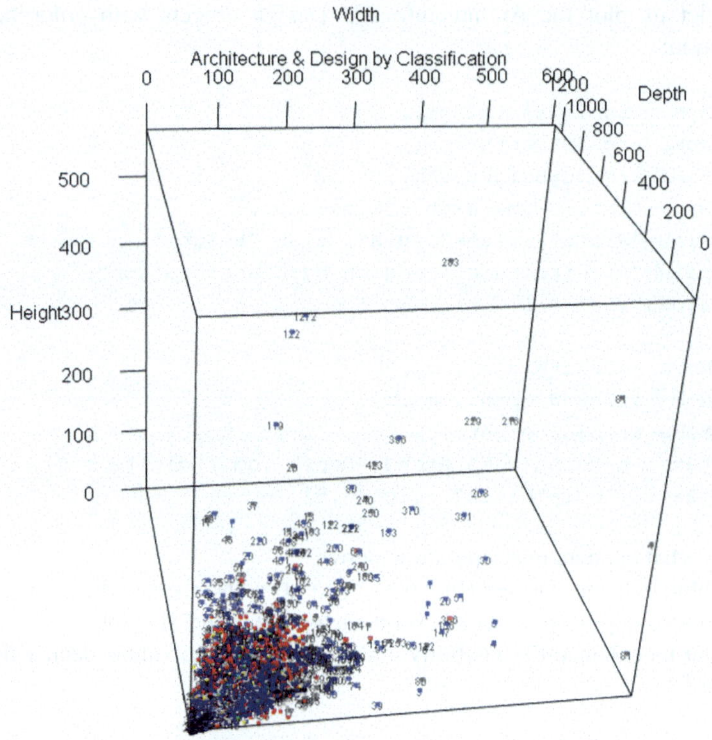

Fig. 6.8 3D scatter, with labels for each data point

```
plot3d(MoMa_arch3d$Width..cm.,
       MoMa_arch3d$Depth..cm.,
       MoMa_arch3d$Height..cm.,
       size=4, col=MoMa_arch3d$class_col,
       xlab="Width", ylab="Depth", zlab="Height",
       main="Architecture & Design/Classification",
       alpha=.4)
text3d(MoMa_arch3d$Width..cm.,
       MoMa_arch3d$Depth..cm.,
       MoMa_arch3d$Height..cm.,
       texts = ifelse(MoMa_arch3d$Depth..cm. > 250,
                      MoMa_arch3d$Classification,
                      ""),
       cex=.8, col="black", alpha=.6)
```

It would be ideal if we could directly read the title of each work from the plot. Unfortunately, the function does not support umlauts and accents. For example, the title in line 2211 is "911 coupé." Fortunately, we can convert it to a string without these special characters, like "911 coupe."

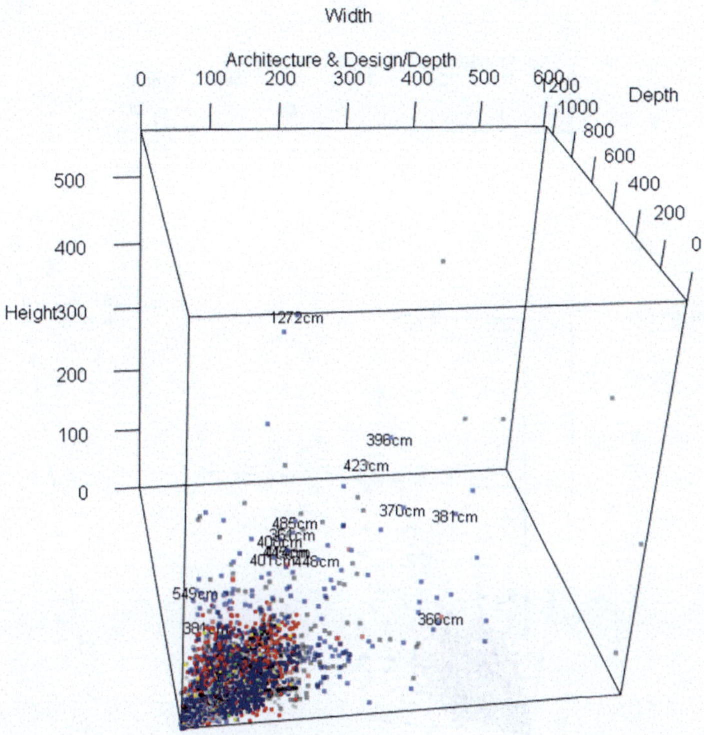

Fig. 6.9 3D scatter, labeled with specific measurements for selected data points

```
MoMa_arch3d$Title[2211]
iconv(MoMa_arch3d$Title[2211], from="UTF-8", to="ASCII//
TRANSLIT")
```

We can use this converted title as the label in our plot:

```
plot3d(MoMa_arch3d$Width..cm.,
       MoMa_arch3d$Depth..cm.,
       MoMa_arch3d$Height..cm.,
       size=4, col=MoMa_arch3d$class_col,
       xlab="Width", ylab="Depth", zlab="Height",
       main="Architecture & Design/Title",
       alpha=.4)
text3d(MoMa_arch3d$Width..cm.,
       MoMa_arch3d$Depth..cm.,
       MoMa_arch3d$Height..cm.,
       texts = ifelse(MoMa_arch3d$Depth..cm. > 275,
                      iconv(MoMa_arch3d$    Title, from="UTF-8",
to="ASCII//TRANSLIT")  ,
                      ""),
       cex=.7, col="black", alpha=.5)
```

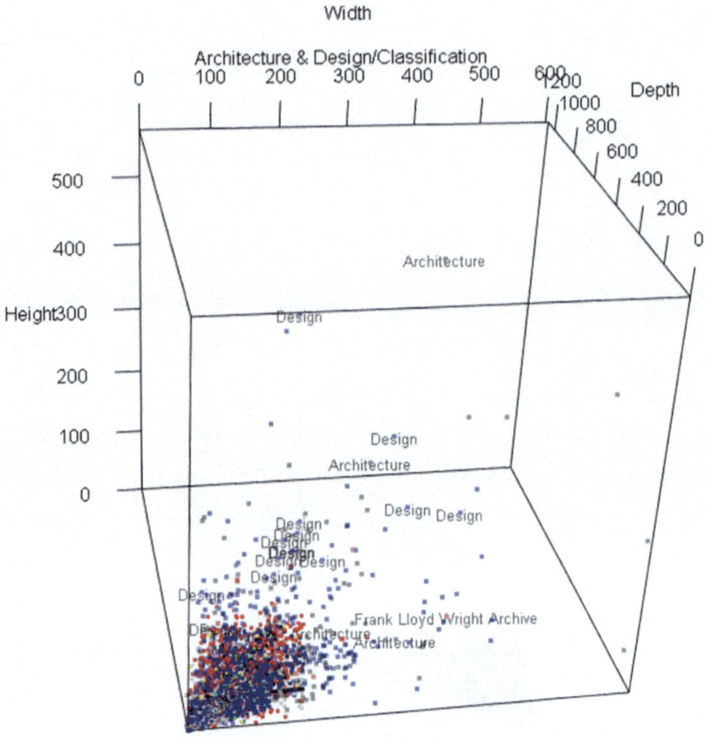

Fig. 6.10 3D scatter plot, with classification labels

This time, if the depth is greater than 275 cm, it writes the converted string; otherwise, it writes nothing (Fig. 6.11).

> Youake some time as the file n "string"mentioned here. While the term "string"is conceptually correct (as it describes a series of characters), in R, the official term used for text data is "character."

Now, let us find out which is the longest object:

```
max(MoMa_arch3d$Depth..cm., na.rm = TRUE)
MoMa_arch3d[which.max(MoMa_arch3d$Depth..cm.),]$Title
```

The longest object is the "Bell-47D1 Helicopter," with a length of 12 m. Another lengthy design object is the "Night Heron" Kayak.

You can also save this plot as an animated GIF (Fig. 6.12):

```
movie3d(spin3d(), duration=10, convert=TRUE, dir=getwd())
```

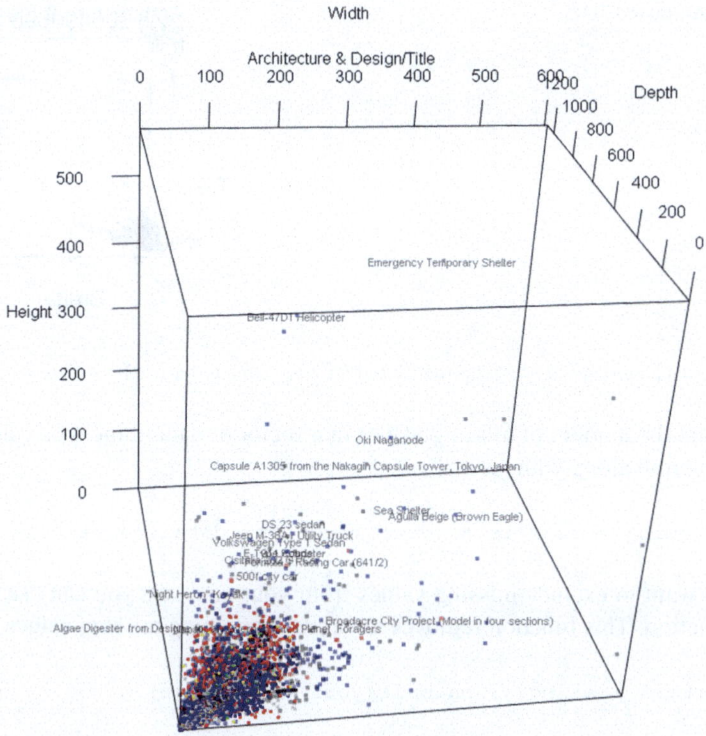

Fig. 6.11 3D scatter plot, with labels for specific objects

Make sure the output window is still open. Additionally, ImageMagick needs to be installed.[3] Alternatively, set "convert" to "FALSE" to obtain individual frames that can be combined in a video editing program.

You can adjust the size of the animation by manually resizing the output window before executing "movie3d" or precisely set the size with:

```
r3dDefaults$windowRect <- c(100, 100, 1000, 1000)
```

Each of these methods offers a unique way to visualize the depth of artworks in a three-dimensional space.

6.6 Dealing with Missing Values

If you want to check if a value is empty in R, you can use "is.na" and "is.not.na." These functions are particularly useful when dealing with missing data. It returns a logical vector indicating whether each element in the vector or column of the data frame is missing or not (`True` or `False`).

[3] https://imagemagick.org/script/download.php

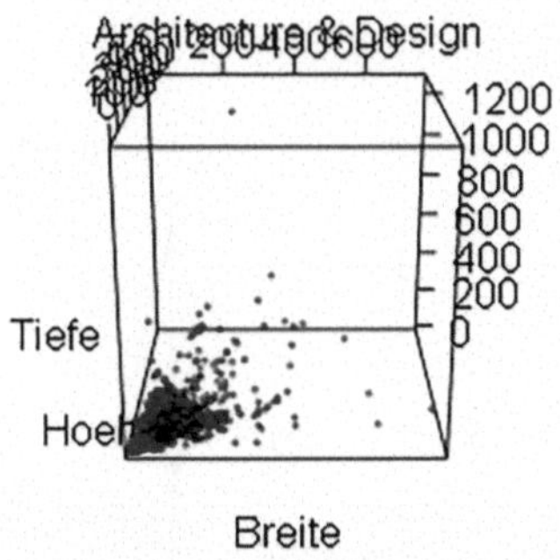

Fig. 6.12 Animated 3D scatter plot

```
missing_values <- is.na(your_vector_or_data_frame)
```

To count the number of missing values in a vector or data frame, you can use the "sum" function along with "is.na."

```
count_missing <- sum(is.na(your_vector_or_data_frame))
```

If you want to exclude missing values from your analysis, you can use the "na.omit" function. This function removes any rows containing missing values.

```
data_without_missing <- na.omit(your_data_frame)
```

You can replace missing values with a specific value using the "ifelse" function. For example, replacing missing values with 0:

```
your_vector_or_data_frame <- ifelse(is.na(your_vector_or_data_
frame), 0, your_vector_or_data_frame)
```

When performing data analysis, it is important to consider the impact of missing values. Functions like "complete.cases" can be used to identify complete cases in a data frame, i.e., rows without any missing values.

```
complete_cases <- your_data_frame[complete.cases(your_data_
frame), ]
```

This can be useful when you want to focus on data sets with complete information.

Understanding and managing missing data is crucial for ensuring the accuracy and reliability of your analyses in R.

> Note: There is a crucial difference between several types of empty data: For example, an empty string ("") is not recognized as NA because there is something in it, namely nothing. In many cultural data, you will also find the string "no data." In data terms, this is of course information, namely a string indicating that no data is available. When it comes to cultural data, we must always keep in mind that it was not originally created for the public but for internal purposes and may have a long history of data acquisition.

6.7 Saving and Loading Objects

Usually, when you start RStudio, the previously saved workspace is restored. If you close the program using "quit session," the objects are saved in the .RData file in the project directory. For additional safety, you can manually save the workspace:

```
save.image(file = "my_work_space.RData")
```

To load the workspace back in the same location:

```
load("my_work_space.RData")
```

If you want to save only a specific object (here, "collection"), creating a CSV file is a good option:

```
write.csv(collection, file = "my_collection.csv")
```

And to import it back:

```
my_collection <- read.csv("my_collection.csv", header=FALSE)
```

> Please refer to the help entry for more details.

For instance, you can import our MoMA data using this command:

```
Artworks <- read.csv("Artworks.csv")
```

Similarly, you can directly import the MoMA file from the web:

```
Artworks <- read.csv("https://github.com/MuseumofModernArt/
collection/raw/main/Artworks.csv?download=")
```

This process may take some time as the file needs to be downloaded (watch for the red dot).

You can save an object or all objects as an R file:

```
# For one object
save(Artworks, file = "backup-MoMA.RData")
# For all objects
save.image("backup-all.RData")
```

6.8 Pie Chart

As described above, MoMA also offers two other data sets that may be of interest. We can download this and import it from there:

```
# Importing additional datasets
MoMADirectorsDepartmentHeads <-
read.csv("MoMADirectorsDepartmentHeads.csv", encoding="UTF-8")
MoMAExhibitions1929to1989 <-
read.csv("MoMAExhibitions1929to1989.csv", encoding="UTF-8")

# Counting the number of directors by gender
gender_counts <- table(MoMADirectorsDepartmentHeads$Gender)

# Creating a data frame with gender counts and calculating the
percentage
unique_directors_gender <- as.data.frame(gender_counts)
unique_directors_gender$perc <-
100/nrow(MoMADirectorsDepartmentHeads) *
unique_directors_gender$Freq

# Creating a pie chart
pie(unique_directors_gender$perc,
    labels <- paste(unique_directors_gender$Var1,
round(unique_directors_gender$perc), "%"),
    main <- "Gender Distribution of MoMA Directors and Department
Heads"
)
```

In this expanded code, we import two additional datasets, "MoMADirectorsDepartmentHeads" and "MoMAExhibitions1929to1989." Then we count the number of directors by gender using the "table()" function and create a data frame "unique_directors_gender" to store gender counts and calculate the

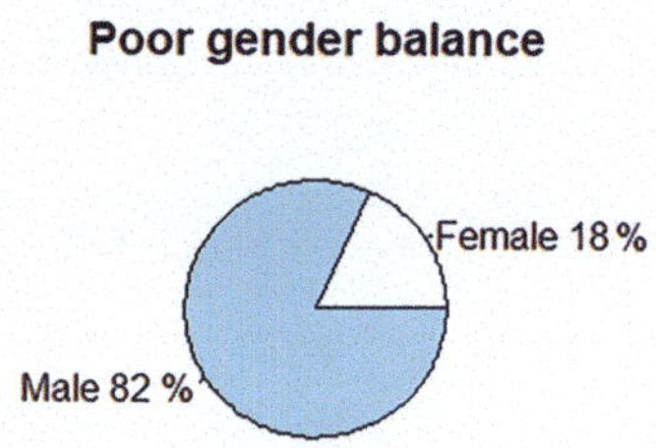

Fig. 6.13 Pie chart showing generder balance

percentage. The "pie()" function is then used to create a pie chart based on the gender distribution.

This pie chart (Fig. 6.13) visually represents the gender distribution of MoMA directors and department heads, providing a quick overview of the proportion of male and female leaders.

6.9 Tips

If you have incorrectly named a column, you can easily copy it to another:

```
Artworks$adate <- Artworks$xdatum
```

If you have accidentally created one column too many, delete it like this:

```
Artworks$xdatum <- NULL
```

> ℹ If you want a better overview of R, look at the Cheat Sheets, especially Base R and R Studio IDE. And an overview of commands in R: https://posit.co/resources/cheatsheets/

Chapter 7
Case Study 1: Picasso from Creation to Acquisition

In the MoMA dataset, we have three important dates: the birth year of the artists, the creation date of the artwork, and the date when it entered the collection. By calculating the differences between these dates, we can determine at what age of the artist the artwork was created and how much time passed before the artwork entered the collection.

To do this, we create columns with the converted dates into numerical and a column to store the calculated differences, which we can then visualize. Here is a walk through this process using Picasso's artworks as an example. This will allow us to quickly see when each of Picasso's works entered the collection. By comparing this information with other artists, we can make qualitative interpretations that require context.

Here is how we can realize this data-driven approach to understanding the museum's holdings and the artists it represents:

1. Data Preparation: Convert the date columns (birth year, creation date, collection date) to a suitable date format if they are not already in one.
2. Calculate Age at Creation: Subtract the artist's birth year from the creation date to determine at what age the artist created the artwork.
3. Calculate Time to Collection: Subtract the creation date from the collection date to find out how much time passed before the artwork entered the collection.
4. Subset Data for Picasso: Create a subset of the dataset that includes only Picasso's artworks.
5. Visualization: Visualize the calculated ages at creation and time to collection for Picasso's artworks. This could be done using scatter plots, histograms, or any suitable visualization method.
6. Comparative Analysis: To provide qualitative context, compare Picasso's artworks with those of other artists. Are there patterns or differences in terms of when their works entered the collection? This might lead to interesting insights into the museum's acquisition practices or the artists' careers.

H. Klinke, *Cultural Data Science*, Quantitative Methods in the Humanities and Social Sciences, https://doi.org/10.1007/978-3-031-88130-5_7

This chart displays the time from creation to inclusion in the MoMA collection.

The prerequisite is that the columns “cdate” (creation date) and “adate” (collection date) exist in the dataset.

First, we create a Picasso subset called “MoMa_Picasso.” We can do this using the following command:

```
MoMa_Picasso <- subset(Artworks, grepl("Picasso",
Artworks$Artist))
```

While a subset with the double equals sign (“==”) would only extract entries where the artist exactly matches “Pablo Picasso,” using “grepl” helps us identify entries that contain the word “Picasso.” This way, we can also find works in which he was involved.

Next, we create a new table with “from” and “to” values:

```
picasso_fromto <- data.frame(from = MoMa_Picasso$cdate, to =
MoMa_Picasso$adate)
```

We then remove any rows with missing values (NAs):

```
picasso_fromto <- na.omit(picasso_fromto).
```

Now, we sort the list:

```
picasso_fromto_ordered <-
picasso_fromto[order(picasso_fromto$from),]
```

Finally, we calculate the differences between the “to” and “from” values:

```
picasso_fromto_ordered$diff <- picasso_fromto_ordered$to -
picasso_fromto_ordered$from
```

This process results in the following table (Fig. 7.1):

Then, we create the plot using a loop (Fig. 7.2):

Fig. 7.1 Table showing the years “from” and “to,” along with the calculated difference

from	to	diff
1901	1998	97
1904	1956	52
1904	1993	89
1905	1985	80
1905	1934	29

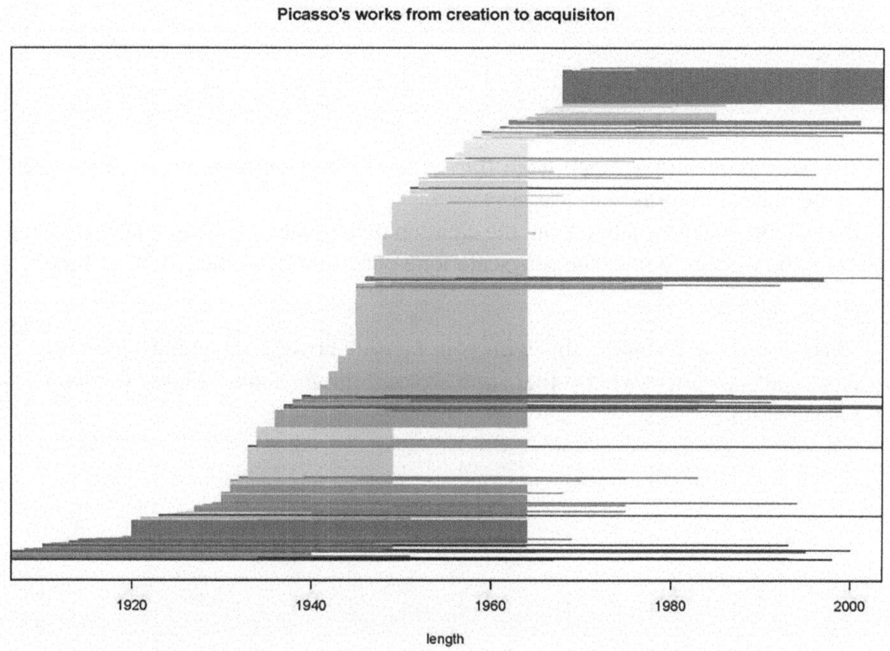

Fig. 7.2 Timeline visualization, highlighting the length of time before they were acquired

```
dev.off()


maxlength <- max(picasso_fromto_ordered$diff, na.rm = TRUE)


for (i in 1:nrow(picasso_fromto_ordered)) {


  plot(x = c(picasso_fromto_ordered$from[i],
picasso_fromto_ordered$to[i]),
       y = c(i, i),
       type = "l",
       xlim = c(1910, 2000),
       ylim = c(0, nrow(picasso_fromto_ordered)),
       xlab = "Length",
       ylab = "",
       yaxt = 'n',
       lwd = 1,
       main = "Picasso's works from creation to acquisition",
       col = gray(1 - picasso_fromto_ordered$diff[i] / maxlength))
```

```
  par(new = TRUE)


}
```

The grayscale values range from 0 to 1, and "maxlength" is required to determine the maximum grayscale value.

In the loop, we draw a line from the creation date to the acquisition date, stacking them on top of each other. The grayscale tone corresponds to the length of time.

> ℹ Now you can try doing this with your favorite artist, compare different artists, and consider what other line colors might make sense for your visualizations.

Chapter 8
Subdividing and Combining Datasets

We saw in the previous case study that subdivision into subsets as well as comparison with other subsets makes sense to relate subareas to one another. Let us start by taking a closer look at how to create subsets and work with them.

8.1 Handling Subsets

You might be familiar with set theory from school—what belongs to a set, what does not, and what two sets have in common.

When creating a subset of *Artworks*, you might not want to include all columns, especially when you only need specific columns for the next step or for clarity.

In other words, if you want to limit the set not only in terms of rows but also columns, look for the "select" parameter in "subset."

```
subset <- subset(Artworks, Artworks$Date == "1901", select =
c(Date, DateAcquired))
```

This creates a subset with the year 1901 but only includes the Date and DateAcquired columns.

8.2 Subset with Partial Match

If you want to create a subset with a specific term, simply specify the string after the double equal sign.

H. Klinke, *Cultural Data Science*, Quantitative Methods in the Humanities and Social Sciences, https://doi.org/10.1007/978-3-031-88130-5_8

```
subset_picasso <- subset(Artworks, Artworks$Artist == "Pablo
Picasso")
```

This results in 1309 objects.

However, if you want to find this string anywhere in the variable (i.e., when Picasso is involved), you can use "grep" or "grepl."

Within the function `subset`, you need a function that provides a logical statement (true/false), which is what "grepl" does.

```
subset_picasso <- subset(Artworks, grepl("Picasso",
Artworks$Artist))
```

This results in 1322 objects.

8.3 Comparison with a List

If you want to compare not just with one string but with multiple strings, you can use a list. Combine these strings with "c()" and use "%in%" as the operator:

```
subset <- subset(Artworks, Artworks$Artist %in% c("Pablo
Picasso", "Max Beckmann", "George Braques"))
```

Suppose you want to create a subset based on a list of names that should be included in the subset. In that case, you can first create the list:

```
search_terms <- c("Picasso", "Braque")
subset <- subset(Artworks, grepl(paste(search_terms, collapse =
"|"), Artworks$Artist))
```

This results in 1499 objects.

> `collapse ="|"` in the paste() function is used to concatenate the elements of the search_terms vector into a single string, with each element separated by the "|" character, which is a logical OR operator.

If you want to create a subset based on a column from another table, you can use that column as a list for comparison. Let us assume we have a data frame called "data_frame," and its column "list_names" contains a list of names. We want to extract artworks by those artists from "Artworks." You can do it like this:

```
Artworks_artists <- subset(Artworks, Artworks$Artist %in% data_
frame$ list_names)
```

Alternatively, you can use the OR operator:

```
artist_czechoslovakian <- subset(Artworks, Artworks$Nationality
== "(Czech)" |
                                     Artworks$Nationality ==
"(Czechoslovakian)" |
                                     Artworks$Nationality ==
"(Slovak)")
```

8.4 Excluding Objects

If you want to exclude specific objects, you can use the subset function with the negation of the operator (exclamation mark). So, instead of.

```
Dataset_new <- subset(dataset, vector %in% c('A','B','C'))
```

you would use:

```
Dataset_new <- subset(dataset, !(vector %in% c('A','B','C')))
```

For example, if you want all nationalities except certain ones, where `dataset` is the entire dataset (Artworks), `vector` is the column to be compared (Artworks$nationality), and "A," "B," "C" are the nationalities ("United States," "France," etc.), then Dataset_new would be the subset. By comparing this subset with the dataset containing USA/France and/or with the entire dataset, you already have an interesting result.

8.5 Multiple Conditions

Suppose you want to include all German female artists in a subset. In that case, you need two conditions:

1. Nationality is German, and.
2. Gender is Female.

You connect both conditions with a logical AND operator "&."

```
Artworks_de_fe <- subset(Artworks, Artworks$Nationality ==
"(German)" & Artworks$Gender == "(Female)")
```

Now that we have acquired good competence in working with data, let us explore how to represent subsets among themselves and, in comparison, to the complete dataset. We will also delve into dealing with images associated with the data. It is going to be interesting!

8.5.1 Set Theory

Here is a summary of set operations (Fig. 8.1):

You can use these operations to analyze the commonalities and differences between datasets.

If you want to find common values in two tables, you can use the "intersect" function. For example, to compare two lists of names:

```
intersect(Artworks$Artist, data_frame$list_names)
```

This will give you all the names that appear in both columns. However, if you want to extract all artworks by artists in the common list, you can do it like this (as mentioned earlier):

```
Artworks_new <- subset(Artworks, Artworks$Artist %in% data_
frame$list_names)
```

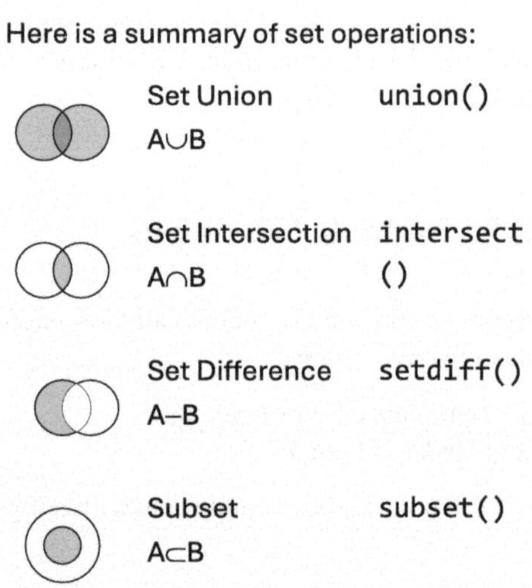

Fig. 8.1 Set operations

8.6 Common Values in Two Tables

Suppose you want to compare the "ConstituentID" column in "Exhibitions" with the one in "Artworks." In that case, you can use "intersect." This will give you a list of shared "ConstituentIDs."

> ℹ The "ConstituentID" is a unique identifier assigned to each artist, ensuring that each artist in the dataset can be distinctly identified.

To use this in "subset," you need a list of logical elements (i.e., True/False), which you can obtain with "is.element":

```
artist_exh <- subset(
      Artworks,
      is.element(Artworks$ConstituentID,
MoMAExhibitions1929to1989$ConstituentID)
)
```

This will result in the intersection of both datasets, similar to "%in%."

> ℹ By the way, "outersect" gives you artists whose works were never part of exhibitions. It represents the difference between two sets.

The union operation is simply called "union."

If you need to add columns from another table where rows correspond to an ID, you can use "merge."

8.6.1 Implementation in R

In Base R, you have access to the following functions:

- "union" returns every row that appears in one or more datasets, only once if it appears multiple times.
- "intersect" returns only the rows that appear in both datasets.
- "setdiff" returns the rows that appear in the first dataset but not in the second.

With the `dplyr` package, you have various join functions available. These operations help you work with and analyze datasets effectively (Fig. 8.2):

ID	X1
1	a1
2	a2

ID	X2
1	b1
2	b2

inner_join

ID	X1	X2
2	a2	b1

left_join

ID	X1	X2
1	a1	NA
2	a2	b1

right_join

ID	X1	X2
2	a2	b1
3	NA	b2

full_join

ID	X1	X2
1	a1	NA
2	a2	b1
3	NA	b2

semi_join

ID	X1
2	a2

anti_join

ID	X1
1	a1

Fig. 8.2 Various types of joins, showing the resulting merged datasets

Chapter 9
Case Study 2: Photography Aspect Ratios

In this second case study, we will explore the aspect ratios of photographs in the MoMA dataset. By analyzing the width and height measurements of these photographs, we can determine the prevalent aspect ratios. Here is how we can achieve this:

We start by assuming that the "Artworks" dataset exists. First, we create a subset containing only photographs:

```
Artworks_Foto <- subset(Artworks, Artworks$Department ==
"Photography")
```

Next, we plot all the aspect ratios with a gray marker on a black background to visualize them:

```
maxdim <- 25
par(bg = 'black')
plot(Artworks_Foto$Width..cm., Artworks_Foto$Height..cm.,
pch = 3,
     main = "Dimensions of Photography", xlab = "Width (cm)",
ylab = "Height (cm)",
     ylim = range(4:maxdim), xlim = range(4:maxdim),
     col = "lightgray", col.lab = "darkgray", col.main =
"darkgray", col.axis = "darkgray",
     bty = "L")
```

To differentiate objects with "Portrait" or "Landscape" in their titles, we create subsets for each category and plot them in distinct colors (Fig. 9.1):

H. Klinke, *Cultural Data Science*, Quantitative Methods in the Humanities and Social Sciences, https://doi.org/10.1007/978-3-031-88130-5_9

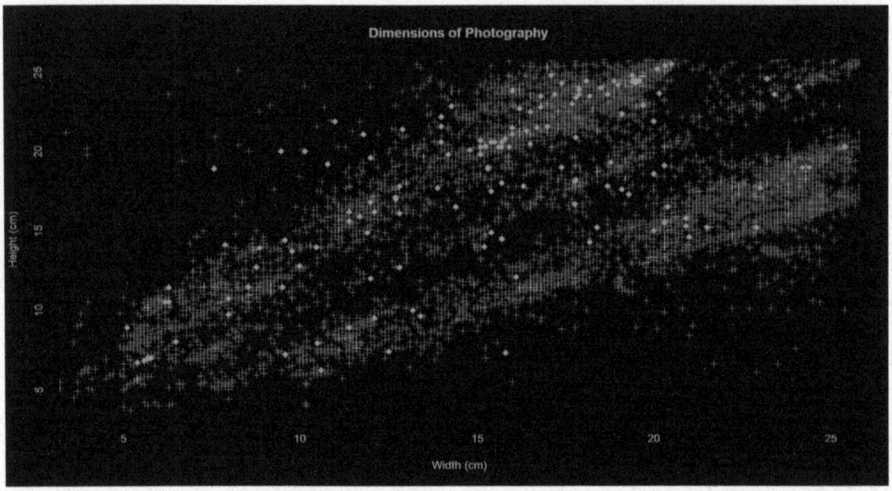

Fig. 9.1 Scatter plot showing the dimensions of photography works in the MoMA collection

```
Artworks_Foto_p <- Artworks_Foto[grep("Portrait", Artworks_Foto$
Title),]
Artworks_Foto_l <- Artworks_Foto[grep("Landscape", Artworks_Foto$
Title),]

par(new = TRUE)


plot(Artworks_Foto_p$Width..cm., Artworks_Foto_p$Height..cm.,
pch = 19,
     main = "Dimensions of Photography", xlab = "Width (cm)",
ylab = "Height (cm)",
     ylim = range(4:maxdim), xlim = range(4:maxdim),
     col = "green", col.lab = "darkgray", col.main = "darkgray",
col.axis = "darkgray")

par(new = TRUE)
plot(Artworks_Foto_l $Width..cm., Artworks_Foto_l$Height..cm.,
pch = 19,
     main = "Dimensions of Photography", xlab = "Width (cm)",
ylab = "Height (cm)",
     ylim = range(4:maxdim), xlim = range(4:maxdim),
     col = "blue", col.lab = "darkgray", col.main = "darkgray",
col.axis = "darkgray")
```

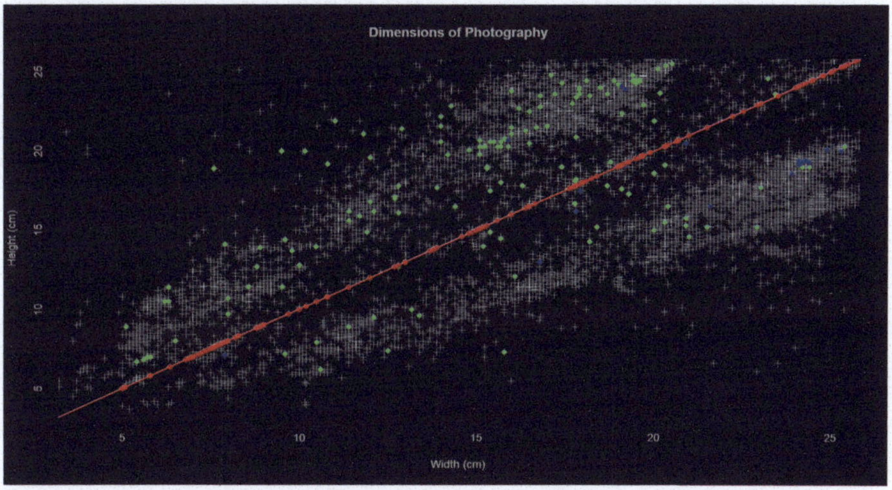

Fig. 9.2 Scatter plot with a linear regression line

Now, let us draw lines for objects with specific aspect ratios. We can either use objects with the same aspect ratios or draw lines for specific aspect ratios like squares or the "1:sqrt(2)" ratio (DIN A4) (Fig. 9.2):

```
Artworks_Foto_ee <- subset(Artworks_Foto, Artworks_Foto$Width..
cm. == Artworks_Foto$Height..cm.)
par(new = TRUE)

# All with _ee in red
plot(Artworks_Foto_ee$Width..cm., Artworks_Foto_ee$Height..cm.,
pch = 19,
     main = "Dimensions of Photography", xlab = "Width (cm)",
ylab = "Height (cm)",
     ylim = range(4:maxdim), xlim = range(4:maxdim), type = "b",
     col = rgb(1, 0, 0, 0.5), col.lab = "darkgray", col.main =
"darkgray", col.axis = "darkgray",
     lwd = 2, bty = "L")

# Or for "1:sqrt(2)"
par(new = TRUE)

plot(c(0, 100 * sqrt(2)), c(0, 100), pch = 19,
     main = "Dimensions of Photography", xlab = "Width (cm)",
ylab = "Height (cm)",
     ylim = range(4:maxdim), xlim = range(4:maxdim), type = "b",
     col = rgb(0, 1, 0, 0.5), col.lab = "darkgray", col.main =
"darkgray", col.axis = "darkgray",
     lwd = 2, bty = "L")
```

We can add a legend to the plot to indicate different aspect ratios and categories:

```
# Legend for aspect ratios
legend("topleft", title = "Aspect Ratios",
       legend = c("Golden Ratio", "3:2", "1:sqrt(2)", "4:3",
"Square"),
       col = c("orange", rgb(0, 0, 1), rgb(0, 1, 0), "yellow",
rgb(1, 0, 0)),
       text.col = "darkgray",
       lty = 1, bty = "n", xpd = TRUE)

# Legend for categories
legend("bottomright", legend = c("Objects", "'Portrait'",
"'Landscape'"), pch = c(3, 19, 19),
       col = c("gray", "green", "blue"), text.col = "darkgray",
       bty = "n", xpd = TRUE)
```

Additionally, you can mark specific aspect ratios, like the A4 paper size:

```
# indicate A4 paper size
symbols(x = 21.0,
        y = 29.7,
        circles = 0.75,
        add = TRUE, inches = FALSE,
        bg = rgb(0, 1, 0, 0.5)
)
text(21.0, 29.7, labels = "A4")
```

By increasing maxdim to 43 and adding.

```
png(filename = ".png", width = 1500, height = 1500, units = "px",
res = 150)
```

at the beginning of your script and.

```
dev.off()
```

at the end, you can save the output directly as a PNG file (Fig. 9.3).

This series of steps allows you to visualize and analyze the aspect ratios of photographs in the MoMA dataset, distinguishing between distinct categories (e.g., “Objects,” “Portrait,” and “Landscape”) and specific aspect ratios (e.g., Golden Ratio, 3:2, 4:3).

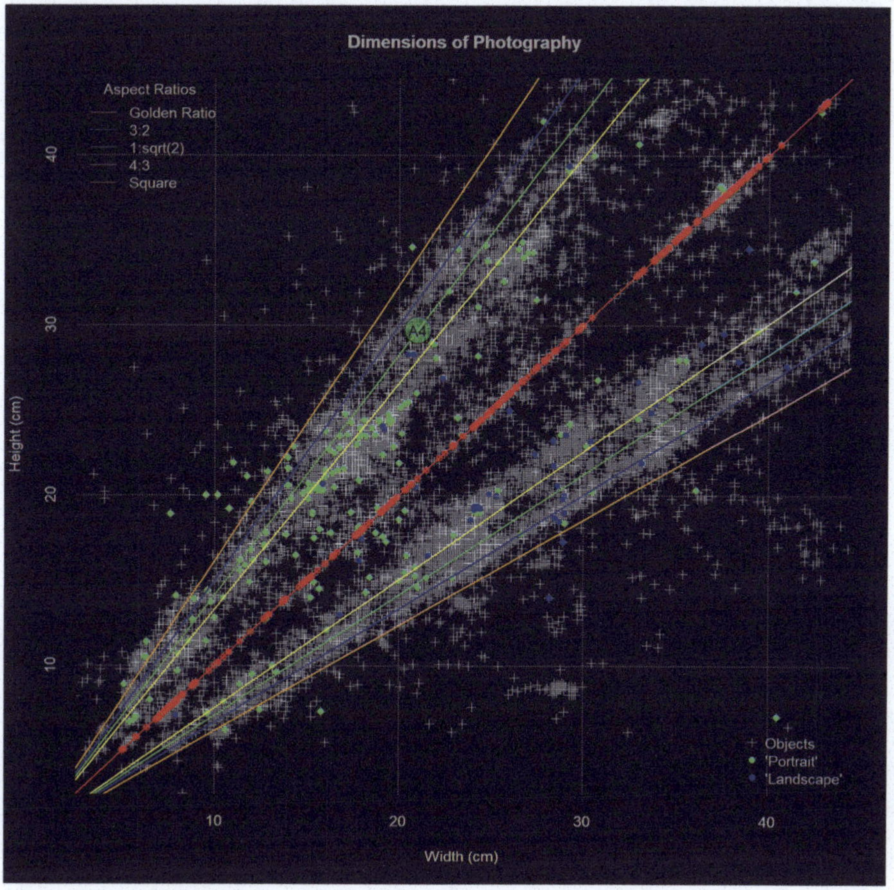

Fig. 9.3 Scatter plot of photography dimensions, with aspect ratio lines

The visualization makes clear that not all objects containing the terms "portrait" or "landscape" in their title necessarily correspond to the respective portrait or landscape formats. It also reveals that the 4:3 aspect ratio appears to be a common, though not the most frequent, format for photography in the collection.

Chapter 10
Case Study 3: Overview of the MNBAQ Dataset

In this third case study, we will explore the *Collections du Musée national des beaux-arts du Québec* (MNBAQ) dataset, similar to what we did with the MoMA dataset.[1] We will begin by loading the MNBAQ dataset into your R environment and then analyze its structure and contents. Here are the steps:

First, download the CSV-file into a folder and import it into R:

```
mnbaq <- read.csv("mnbaqcollections23septembre2016.csv",
                  encoding="UTF-8", dec=".")
```

We use 'dec="."' to specify that the decimal separator is a period. While in Europe, the period is used, in the United States and some regions of Canada, a comma is used. The default is a period, but specifying it here is good practice. This dataset contains 40,004 objects with 16 attributes, and all fields are of type character.

Now, let us explore the dataset by examining some key attributes:

```
nrow(mnbaq)   # Number of rows
ncol(mnbaq)   # Number of columns
summary(mnbaq)   # Summary of the dataset
```

[1] https://ouvert.canada.ca/data/dataset/e42d4c32-3dec-4d73-855f-0fa335006d3d.

H. Klinke, *Cultural Data Science*, Quantitative Methods in the Humanities and Social Sciences, https://doi.org/10.1007/978-3-031-88130-5_10

We can analyze the distribution of objects across different departments and collections:

```
# Departments
sort(table(mnbaq$Département.Collection), decreasing = TRUE)
pie(sort(table(mnbaq$Département.Collection), decreasing = TRUE))

# Collections
sort(table(mnbaq$Collection), decreasing = TRUE)
pie(sort(table(mnbaq$Collection), decreasing = TRUE))
```

These code snippets show that there are eight departments, with the "Art moderne" department from the first half of the twentieth century being the largest with over 10,000 objects (Fig. 10.1). There are also four collections, with over 35,000 objects in the "Collection permanente."

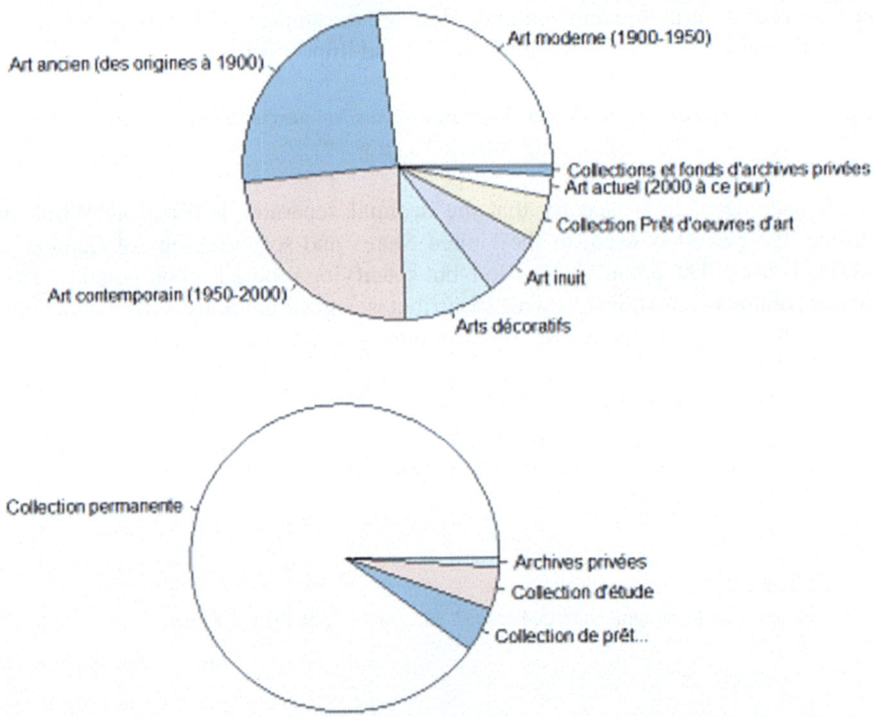

Fig. 10.1 Pie charts illustrating the distribution of artworks in different categories

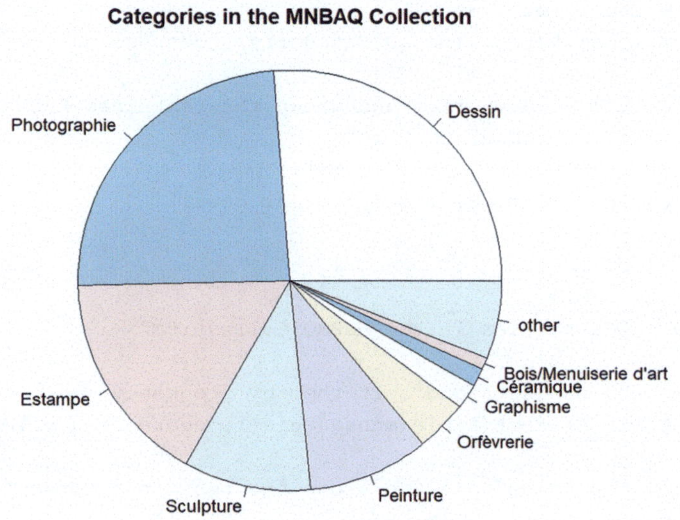

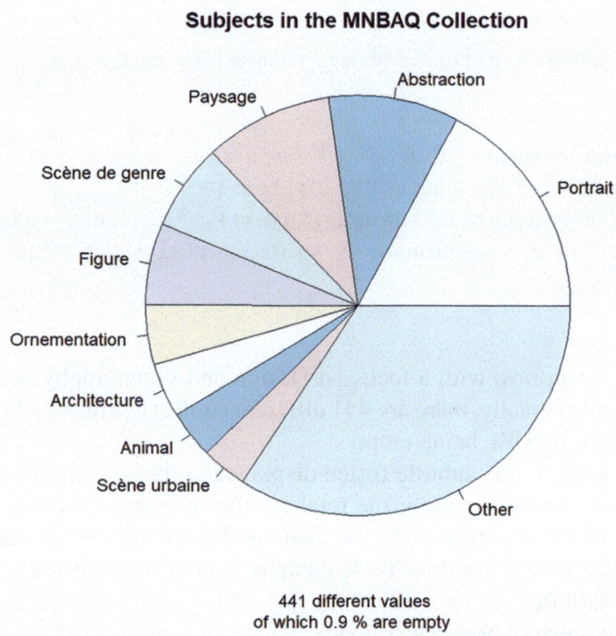

Fig. 10.2 Pie charts showing the distribution of categories and subjects in the MNBAQ collection

Next, we can explore the distribution of objects across distinct categories and subjects (Fig. 10.2). Since some of the categories and subjects are too infrequent to be displayed clearly, we summarize them into a single group labeled "other" to simplify the visualization and improve readability:

```
# Prepare the categories
cat_table <- sort(table(mnbaq$Catégorie), decreasing = TRUE)

# Keep the top 9 categories and group the rest into "other"
top9 <- head(cat_table, 9)
other_count <- sum(cat_table[-(1:9)])
new_cat_table <- c(top9, other = other_count)

# Plot the pie chart
pie(new_cat_table,
    main = "Categories in the MNBAQ Collection")

# Prepare the subjects and sort them by frequency
subject_table <- sort(table(mnbaq$Sujet), decreasing = TRUE)

# Keep the top 9 subjects and group the rest into "Other"
top9 <- head(subject_table, 9)
other_count <- sum(subject_table[-(1:9)])  # Sum of all smaller
categories
new_subject_table <- c(top9, Other = other_count)

# Plot the pie chart
pie(new_subject_table,
    main = "Subjects in the MNBAQ Collection",
    sub = paste(length(table(mnbaq$Sujet)), "different values\nof
which", round(100 / nrow(mnbaq) * sum(mnbaq$Sujet == ""), 1), "%
are empty "))
)
```

There are 44 categories, with a focus on Design and Photography, and none of them are empty. Additionally, there are 441 different subjects, primarily Portrait and Abstraction, with only 0.9% being empty.

In the "pie" function, the subtitle (often displayed at the bottom of a pie chart) includes two pieces of information: the total number of subjects (Sujets) and the percentage of empty values relative to the total number of rows in the dataset. The "paste" function is used to combine these elements into a single string that will be displayed as the subtitle.

Here is a breakdown of how it was done:

1. Creating frequency tables.
2. Grouping infrequent values into a single "other" category.
3. Visualizing the top nine values along with the "other" category in pie charts.
4. Adding a subtitle for the "Sujet" pie chart to summarize the total unique subjects and the percentage of empty entries.

Note of the function of "\n." This backslash-n combination represents a newline character, which signifies a line break in text. When "\n" is encountered in a string, it instructs the text to move to the next line when displayed or printed. This is particularly useful for creating well-formatted and readable output, especially when you want to separate content into distinct lines or paragraphs.

In coding, "\n" is a common escape sequence used to represent newline characters, making it easy to structure text and improve its clarity.

We can now analyze the distribution of objects based on artists and titles:

```
# Artists
pie(sort(table(mnbaq$Artiste), decreasing = TRUE)[1:10], main =
"Top 10 artists")

# Titles
pie(sort(table(mnbaq$Titre), decreasing = TRUE)[1:10], main =
"Top 10 titles")
```

The result (Fig. 10.3) shows: The most common artist is *unknown*. Following closely is *Edmond-Joseph Massicotte*, a Canadian illustrator and printmaker, who is associated with an impressive 1677 artworks. The most common titles are "sans titre" (without title), followed by "cuiller à potage" (soup spoon).

The creation dates are mostly in the format `yyyy` (year), but there are also various other date patterns that are not easily convertible.

Here is a way to get additional insights into the collection:

```
pie(sort(table(mnbaq$Lieu.de.production),decreasing=T)[1:10],
main = "Top10 places" )


pie(sort(table(mnbaq$Culture),decreasing=T)[1:10], main = "Top10
cultures" )
sum(mnbaq$Culture=="")

pie(sort(table(mnbaq$Technique),decreasing=T)[1:10], main =
"Top10 Technique")
sum(mnbaq$Technique=="")
```

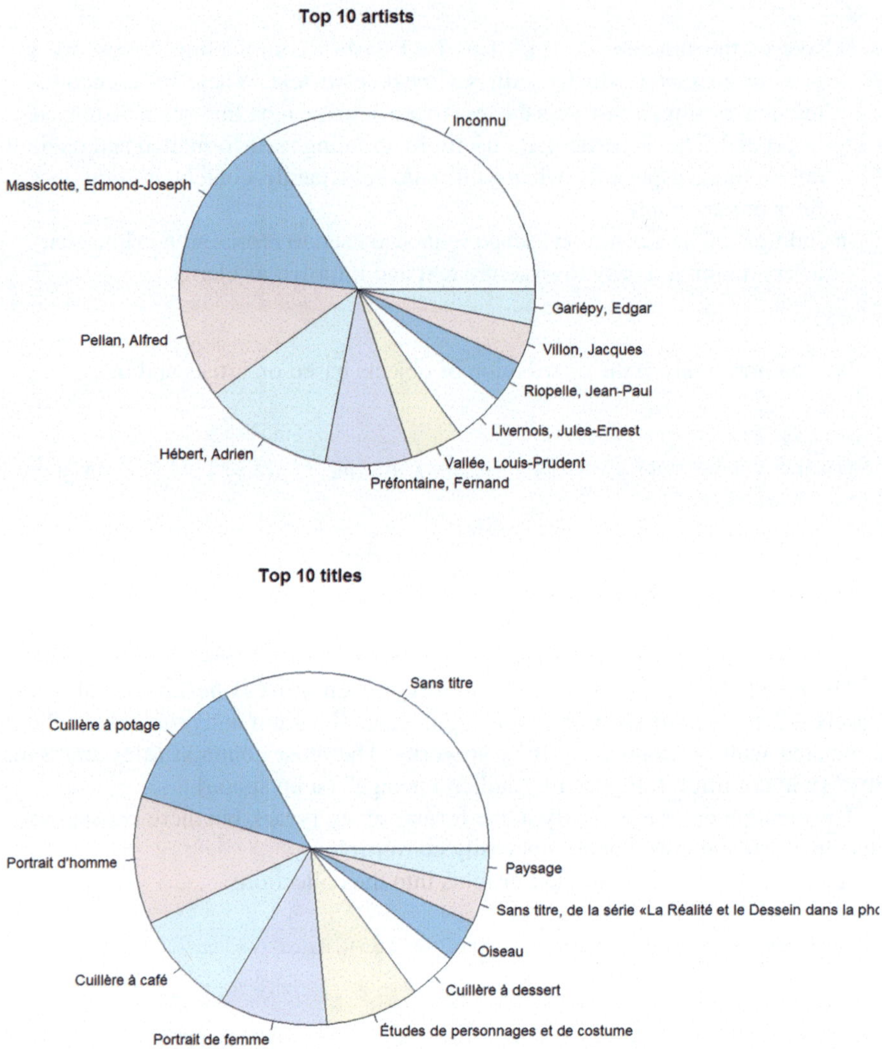

Fig. 10.3 Pie charts showing the top 10 artists and the top 10 titles in the MNBAQ collection

The first pie chart illustrates the top ten places associated with the artworks in the MNBAQ collection, providing insights into the geographical diversity of the collection (Fig. 10.4). Another pie chart displays the top ten cultures represented in the artworks, revealing the cultural diversity within the collection. The third pie chart highlights the most common artistic techniques used, showcasing the Similarly, a pie chart displays the top ten artistic techniques or methods used in the collection, offering insights into the various artistic practices within the MNBAQ collection.

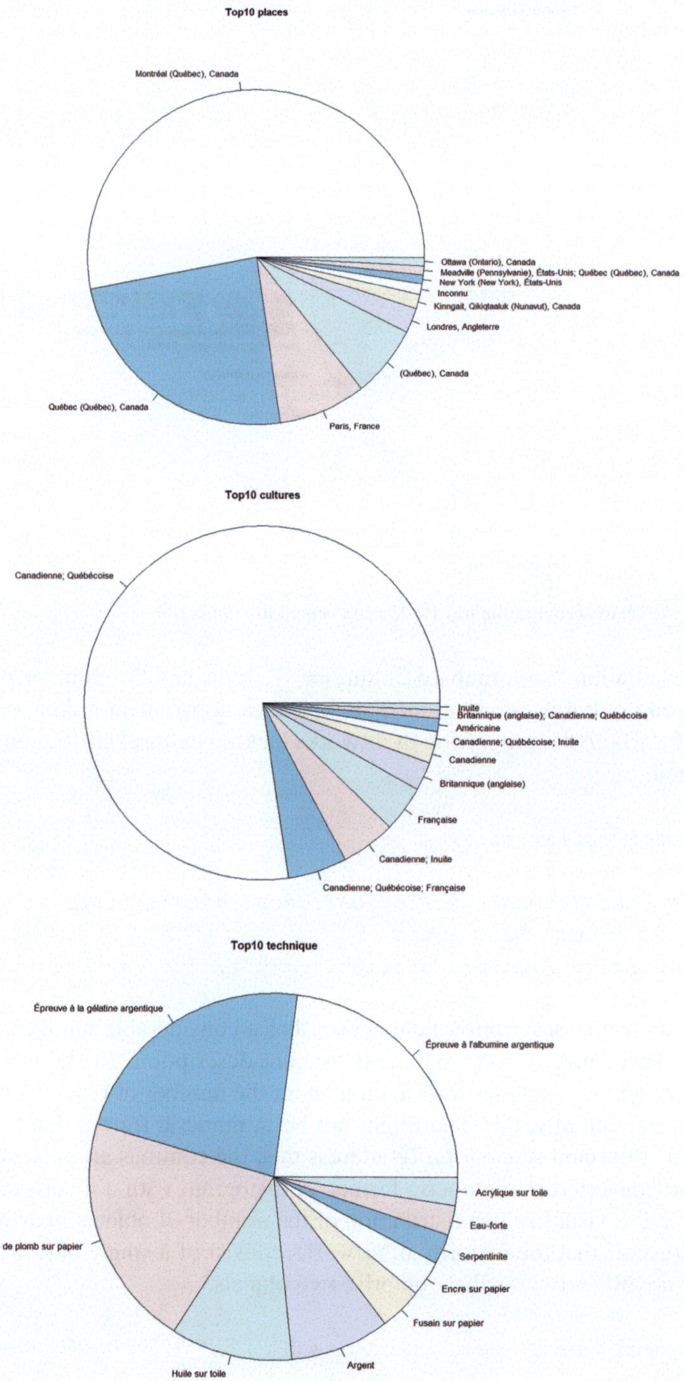

Fig. 10.4 Pie charts showing the top 10 places, cultures and techniques in the collection

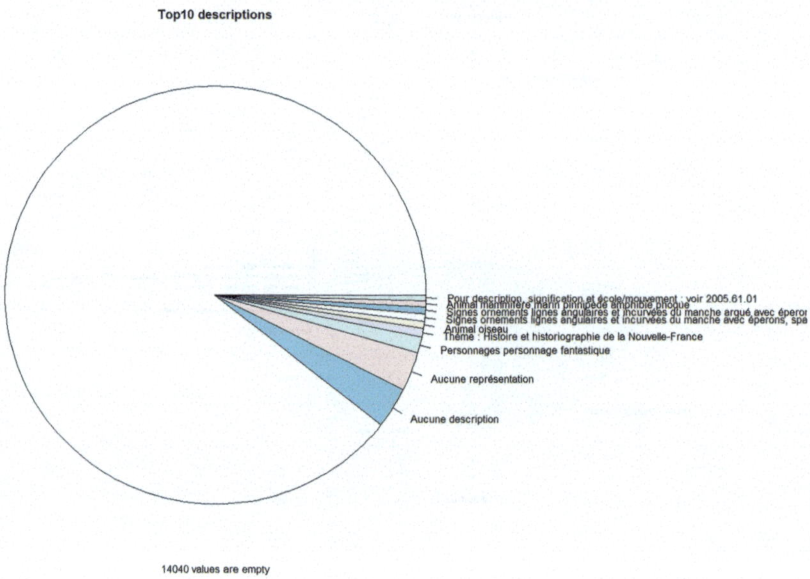

Fig. 10.5 Pie chart showing the top 10 descriptions in the collection

The calculation "sum(mnbaq$Culture==``)" indicates the number of artworks with missing cultural information. This becomes apparent in a data visualization (Fig. 10.5), where the proportion of artworks lacking cultural attribution is visually represented.

```
head(mnbaq$Description)
tail(mnbaq$Description)
pie(sort(table(mnbaq$Description),decreasing=T)[1:10],
    main = "Top10 Description",
    sub=paste(sum(mnbaq$Description==""), "values are empty"))
```

Analysis of the description field shows that a considerable number of descriptions are either empty or contain the text "Aucune description" (without description).

The dataset also includes information about the number of objects that make up each artwork. Initially, this data might not be in numeric format due to the use of commas as thousand separators. To address this, the commas are removed, and the values are converted into numeric format. A histogram with a logarithmic scale is then created to visualize the distribution of the number of objects in artworks. This analysis reveals that the majority of artworks consist of a single object (Fig. 10.6), but there are 505 artworks that comprise two objects.

```
mnbaq$Nombre.objets<-as.numeric(gsub(",", "", mnbaq$Nombre.
objets))
```

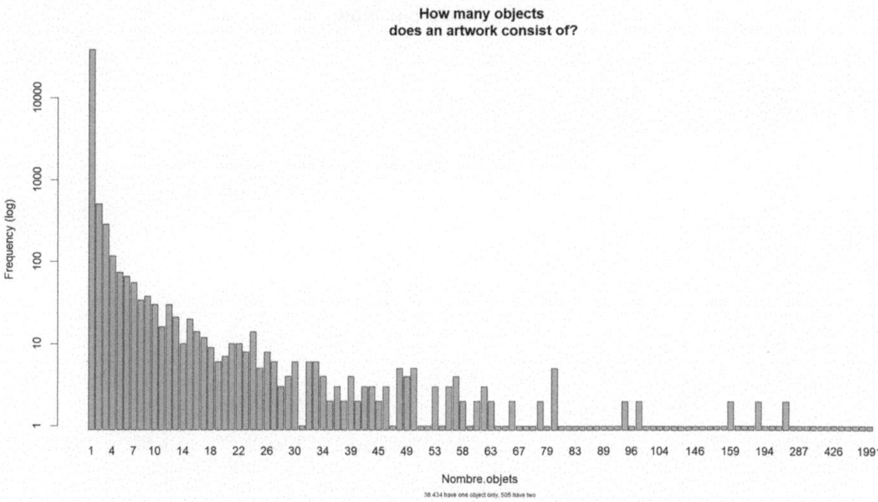

Fig. 10.6 Histogram showing the distribution of the number of objects each artwork consists of

```
dev.off()
barplot(table( mnbaq$Nombre.objets ),
        log = "y", xlab="Nombre.objets", ylab="Frequency (log)",
        main="How many objects\n does an artwork consist of?",
        cex.sub=.5,
        sub = "38.434 have one object only, 505 have two")
```

To understand the field "update date," we can create a histogram:

```
# Data Update Date
hist(as.Date(mnbaq$Date.de.mise.à.jour), breaks = "months",
freq = TRUE,
     main = "Date of Data Update", xlab = "Date",
     sub = "All data were updated in 2015/16")
```

In this analysis (Fig. 10.7), it is evident that all the data entries or updates in the dataset occurred within a few months leading up to the publication date.

The dimensions are more complicated. The MoMA data is handy since each dimension has its own column. Here, in the MNBQ data set much information are separated from each other with an "x," but often not. In a first draft, we do it "quick and dirty" and initially just focus on the largest and most easily convertible group and plot these two dimensions:

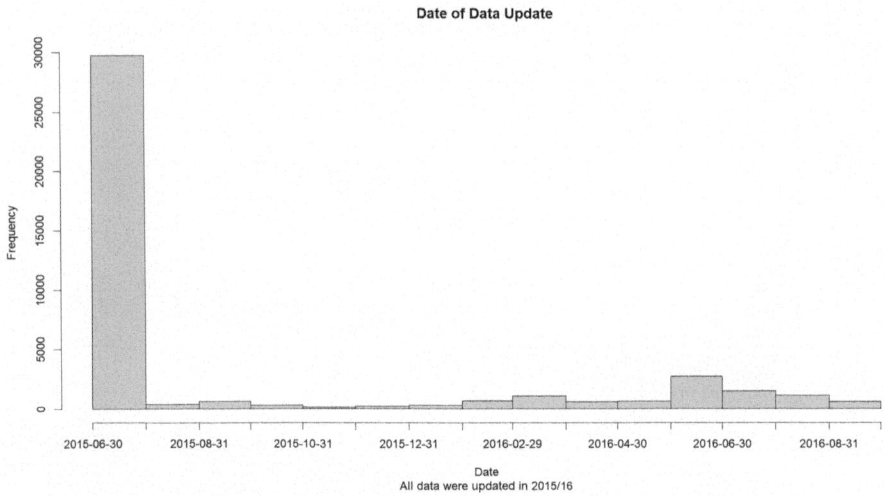

Fig. 10.7 Histogram showing the frequency of data updates over time

```
# Dimensions
for (i in 1:nrow(mnbaq)) {
    mnbaq$h[i] <- as.numeric(unlist(strsplit(mnbaq$Dimensions[i],
" x "))[1])
    mnbaq$w[i] <- as.numeric(unlist(strsplit(unlist(strsplit(mnba
q$Dimensions[i], " x "))[2], " cm"))[1])
    message(i, ". ", mnbaq$h[i], " x ", mnbaq$w[i])
}

plot(as.numeric(mnbaq$h), as.numeric(mnbaq$w),
     xlim = range(0:1500), ylim = range(0:1000),
     pch = "+", col = rgb(0, 0, 0, 0.5))
```

This code helps us visualize the pattern of dimensions in the MNBAQ dataset (Fig. 10.8), just like we did with the MoMA dataset. By following these steps, we gain insights into the structure and content of the MNBAQ dataset, allowing us to analyze various aspects of the artwork collection.

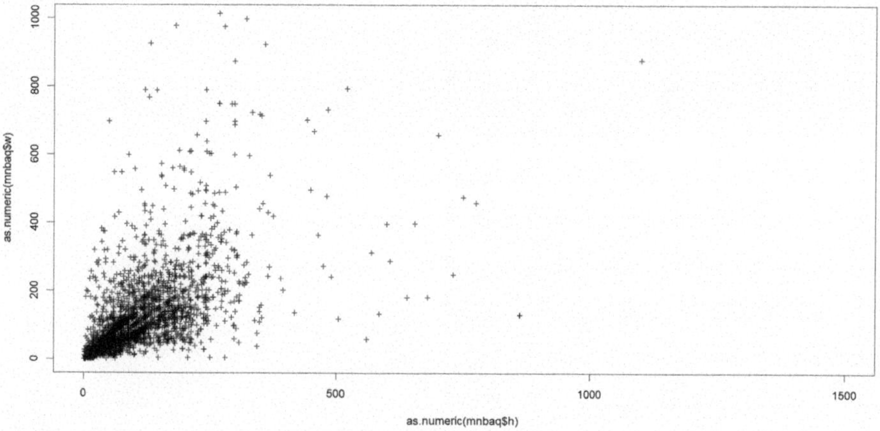

Fig. 10.8 Scatter plot showing the dimensions

Chapter 11
Stacked Bar Plot

Up to this point, we have primarily used XY plots and bar plots for data visualization. When working with cultural data, especially in the context of museums, employing stacked bar plots can be a valuable choice.

A stacked bar plot is a graphical representation that shows how distinct categories or groups are divided into subcategories or components over multiple time periods or variables. Each category is represented as a vertical bar, and within each bar, the subcategories are stacked on top of each other to illustrate their composition and distribution. They are particularly useful when you want to represent multiple categories simultaneously in your visualizations.

Let us explore why stacked bar plots are relevant and effective in the context of cultural data:

In the following R code example, a stacked bar plot is used to visualize the number of objects acquired by different departments of a museum over a range of years (Fig. 11.1).

```
# Define Objects
start_year <- 1929
end_year <- 2019
dep_list <- unique(Artworks$Department)
dep_list_len <- length(dep_list)
data <- matrix(NA, nrow = dep_list_len, ncol = end_year - start_
year + 1)
                                  # From the first year to the last
year, inclusive
colnames(data) <- c(start_year:end_year)
rownames(data) <- dep_list
```


H. Klinke, *Cultural Data Science*, Quantitative Methods in the Humanities and Social Sciences, https://doi.org/10.1007/978-3-031-88130-5_11

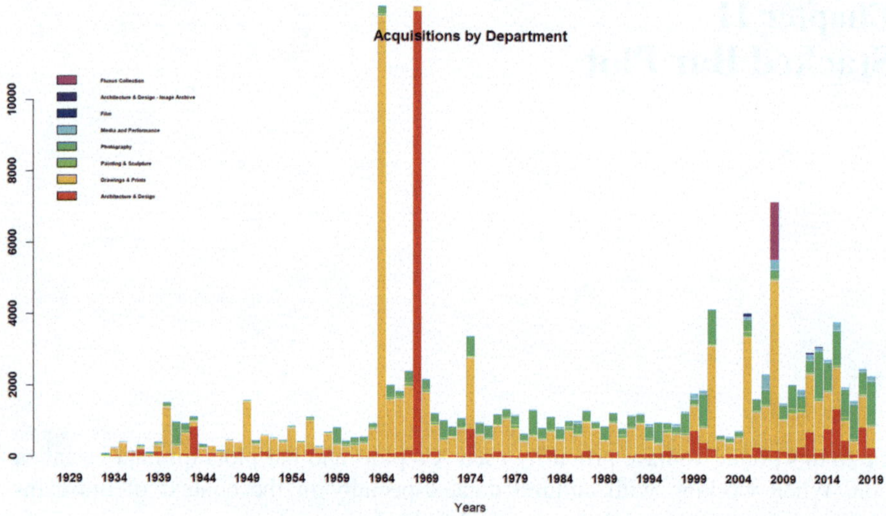

Fig. 11.1 Stacked bar plot illustrating the number of acquisitions by department over the years

```
# Fill the Matrix
for (j in 1:dep_list_len) {
  message(dep_list[j])
  for (i in start_year:end_year) {
    data[j, i - start_year + 1] <- nrow(subset(Artworks,
                Department == dep_list[j] & adate == i))
  }
}

# Plot
barplot(data,
        main = "Acquisitions by Department",
        col = rainbow(dep_list_len),
        border = "white",
        space = 0.04,
        font.axis = 2,
        xlab = "Years",
        ylim = range(0:11000),
        legend.text = rownames(data),
        args.legend = list(x = "topleft", bty = "n", cex = 0.4,
text.font = 2)
        )
```

Let us break down the code step by step:

1. Data Preparation: "start_year" and "end_year" define the start and end years for the analysis. "dep_list" stores unique department names. "dep_list_len" calculates the number of unique departments. "Data" initializes an empty matrix with rows representing departments and columns representing years. This matrix will store the count of objects acquired by each department each year.
2. Matrix Filling: A nested loop is used to iterate through each department and each year in the specified range. "subset(Artworks, Department == dep_list[j] & adate == i)" filters the dataset to select objects that belong to the current department ("dep_list[j]") and were acquired in the current year (""). "nrow(subset(...))" calculates the number of rows in the filtered subset, which corresponds to the count of objects acquired by that department in that year. These counts are filled into the "data" matrix.
3. Plotting: The "bar plot" function is used to create the stacked bar plot. "Data" is the input data containing object counts. Various plot customization options are set, such as titles, colors, borders, and legend placement. The resulting stacked bar plot visually represents how many objects were acquired by each department in each year, allowing for easy comparison and analysis of acquisition patterns over time.

Mind that initially, all cells in the matrix are set to NA. Then, the cells are replaced with either:

- 0 (if no matching rows exist for the department and year).
- A count of acquisitions (if matches are found).

Eventually, any leftover NA values (if present) may indicate departments or years with no data and must be addressed before plotting (e.g., replacing NA with 0 using is.na() and replace).

And here is the same procedure applied to the artworks in MoMA (Fig. 11.2).

Here, we only change the first two lines and replace "Artworks" with "MetObjects," which we have imported first, and the AccessionYear:

```
start_year <- 1870
end_year <- 2019
```

And similarly for the Cleveland Museum of Art (Fig. 11.3):

> ℹ You can find the Cleveland Museum of Art's GitHub repositories at https://github.com/ClevelandMuseumArt. The code snippets for the above diagrams and more can be found on my website for direct integration into your RStudio development environment.

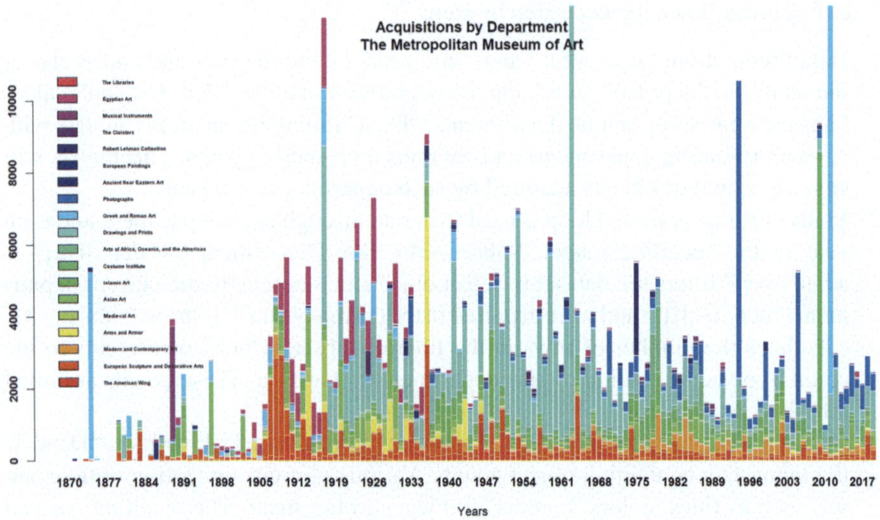

Fig. 11.2 Stacked bar plot showing the number of acquisitions by department at the Metropolitan Museum of Art

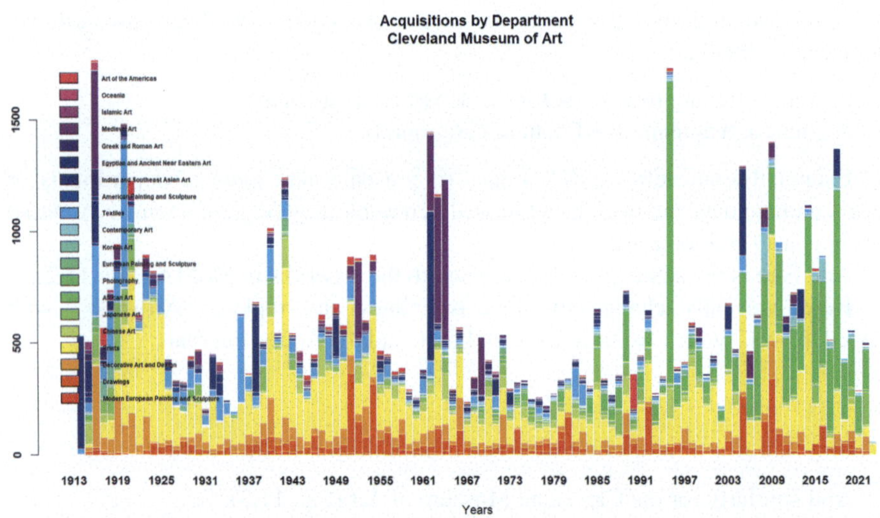

Fig. 11.3 Stacked bar plot showing acquisitions by department at the Cleveland Museum of Art

This type of visualization is valuable for understanding how the distribution of objects varies across different departments and years within a museum's collection. It provides insights into the acquisition trends and priorities of each department over time. Stacked bar plots offer a powerful means of visualizing and interpreting cultural data, enabling you to explore the richness and complexity of collections, exhibitions, and cultural trends.

Chapter 12
Working with Images

In addition to the metadata, we also have the column containing Thumbnail URLs in our dataset. These images can be downloaded and displayed, or further analyzed, to gain insights from the visual content.

The images are available: many artworks have a uniform resource locator (URL) pointing to a small image (thumbnail).

For instance:

```
Artworks$Title[1]
Artworks$ImageURL[1]
```

To achieve this, we could track and download all the thumbnail URLs from the MoMA dataset. However, considering that there are approximately 66,000 individual images, this process takes quite a long time.

To associate the files correctly, the filename corresponds to the `ObjectID`. You can access a specific image like this (`imgs_moma` is the subfolder where we have placed the images):

```
Artworks$ObjectID[1]
my_image_path <- paste0("imgs_moma\\", Artworks$ObjectID[1], ".jpg")
file.show(my_image_path)
```


H. Klinke, *Cultural Data Science*, Quantitative Methods in the Humanities and Social Sciences, https://doi.org/10.1007/978-3-031-88130-5_12

You might wonder why there are two backslashes separating the path components. In programming languages, a backslash (\) is often used as an escape character to represent special characters (e.g., \n for a line break, which we have already encountered). Therefore, a single backslash in a file path could be misinterpreted, so it needs to be escaped itself. Thus, it signifies: "Attention (\), here comes a backslash (\)."

I encourage you to examine the MoMA dataset containing images and explore the other datasets in the Drive folder. Please share any insights you can provide regarding their structure and content.

The focus here is on a subset of photographs that are square in shape, as discussed in Case Study 2. However, you can certainly apply the following steps to a subset of your choice.

Here is how to do it:

1. We assume that you already have a subset of square photographs called "Artworks_Foto_ee" (see code on page 89), which consists of 1127 objects.
2. Our primary interest is in the "Image URL" field of these objects. Please note that many of these fields may be empty, so we need to handle them accordingly.

12.1 Step 1: Download

ⓘYou can skip this step if you have already downloaded the thumbnails as a zip file and extracted them into the subdirectory. However, it is essential to understand how to download images in case you need to do it for a different dataset. In this step, we will create a column called "_thumbOK," which will be accessed later.

The first step of the project is to download the thumbnails via their URLs, but only if the URL is not empty. Here is how we did it:

First, we extracted the URL from the Thumbnail URL column as "thisurl" and generated the filename as "filename," composed of the ObjectID and ".jpg". In essence, if the field is not empty, we download a file from a URL ("thisurl") and save the file as "filename" in binary mode using the "libcurl" method with no output (you can find more about this in the documentation). Additionally, we record whether we successfully downloaded the thumbnail and the filename on the hard drive. We need both pieces of information later—we will only display the image if it exists. Therefore, we need to know the location of the image file.

Now, you might be wondering, what is “thisurl,” what is “filename,” and what is “i”? As you might have guessed, we retrieve the first two from the table, and “i” comes from the loop:

```
subfolder <- "images-squares/"
if (!dir.exists(subfolder)) {
  dir.create(subfolder)
}
for (i in 1:nrow(Artworks_Foto_ee)) {
  thisurl <- Artworks_Foto_ee$ImageURL[i]
  filename <- paste(subfolder, Artworks_Foto_ee$ObjectID[i], ".
jpg", sep = "")
  message(i, "\t", filename)  # So we can see the progress

  if (thisurl != "") {
    download.file(
      thisurl,
      filename,
      mode = "wb", # binary
      method = "libcurl",
      quiet = TRUE
    )
    Artworks_Foto_ee$thumb_OK[i] <- TRUE
    Artworks_Foto_ee$thumb_name[i] <- filename
  } else {
    message("Empty URL")
    Artworks_Foto_ee$thumb_OK[i] <- FALSE
  } # End of if
} # End of for
```

Now, all the thumbnails of our subset are saved on the hard drive in the subdirectory “images-squares.”

12.2 Step 2: Extract Dimensions from Images (Optional)

This step is not absolutely necessary and you could proceed directly to the next step. However, you will soon realize that not all objects listed as square in the table are actually square in their thumbnails. For example, the first one is actually a portrait rectangle. If we want to display only genuinely square images later, we can generate additional data now by measuring the image files.

To use the magick package, you need to install it by going to “Packages” > “Install” and then load it with “library(magick).”

```
library(magick)
# Add thumbnail dimensions
for (i in 1:nrow(Artworks_Foto_ee)) {
  if (Artworks_Foto_ee$thumb_OK[i] == TRUE) {
    image <- image_read(Artworks_Foto_ee$thumb_name[i])
    Artworks_Foto_ee$thumb_width[i] <- image_info(image)$width
    Artworks_Foto_ee$thumb_height[i] <- image_info(image)$height
  }
}
```

Here is what is happening: We have a loop that iterates through all the rows of our subset. Whenever we have a thumbnail available (as indicated by "thumb_OK"), we load the image from "$thumb_name" into the object "image." The function "image_info()" provides us with "$width" and "$height," which we then write into our table as additional columns.

12.3 Step 3: Displaying Images

Now that we have the images saved on our hard drive, we can view them individually. However, it is better to display a subset of images in an organized manner. There are numerous ways to achieve this, and one of the simplest methods is using the "montage()" function from the "magick" package.

> Please note that you might need to have ImageMagick installed on your system. For Windows, you can download and install ImageMagick. For Mac, you will need to install HomeBrew first and then use Terminal for installation.

We could display all the image files in the directory. However, our goal is to display a selection of images, specifically those that are both marked as square in the data and are genuinely square, allowing for a tolerance of 20 pixels.

To achieve this, we will compare the pixel height and width values from Step 2 for equality, with a tolerance of twenty pixels. We use the "all.equal()" function, which simplifies this process. We will loop through the table and create a list of file names:

```
# Create a list
x <- 0
filenames <- NULL
for (i in 1:nrow(Artworks_Foto_ee)) {
  if (all.equal(Artworks_Foto_ee$thumb_width[i],
                Artworks_Foto_ee$thumb_height[i],
                scale = 1,
                tolerance = 20) == TRUE
  ) {
    x <- x + 1
    filenames[x] <- Artworks_Foto_ee$thumb_name[i]

message(i, "\tTRUE\t", x, "\t", Artworks_Foto_ee$thumb_name[i])
  }
}
```

And here is what it might look like:
"images-squares/229413.jpg"

> In case you are wondering about "\t" in the "message" function, it represents a tab character. Analogous to the above-mentioned "\n," it is an escape sequence used to format text output.

Now, let us load all the images from the list of file names into an image list. Just one line of code, really:

```
# Filter out empty or NA filenames
filenames <- filenames[!is.na(filenames) & filenames != ""]
# Create list of images
imagelist <- image_read(filenames)
```

Now, let us display all of them together:

```
# Create a montage
montage <- image_montage(imagelist)
# Display the montage
montage
```

If there is more than one page of images, we can create a montage of the montage:

```
montage_total <- image_montage(montage)
montage_total
```

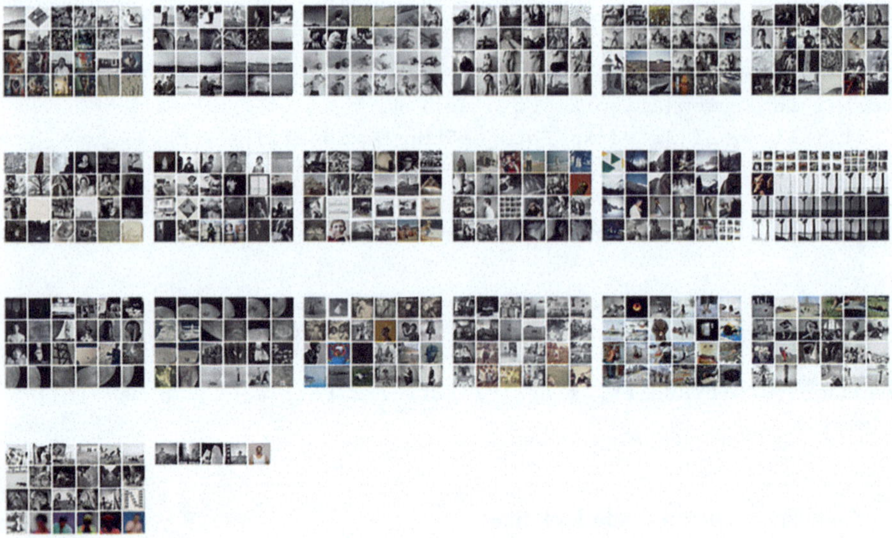

Fig. 12.1 A montage of images representing different works in the collection

You can even save this as an image (Fig. 12.1):

```
image_write(image_convert(montage_total, "png"), "montage_
total.png")
```

12.4 Sorting a List of Images by Brightness (or Other Criteria)

To sort a list of images by brightness (or any other criterion), we first need to calculate the average brightness values. We can easily compute this from the images, which are essentially grids of RGB brightness values (where R, G, and B represent the intensity of red, green, and blue, typically ranging from 0 to 255). The "mean()" function will help us calculate the average, and we will store these values in a new column in our table.

12.5 Step 1: Include the Necessary Libraries

Before we proceed, ensure that you have the "_ee" subset from earlier with columns "$thumb_ok" and "$thumb_name." If you have not already included these libraries, do so now.

```
library(magick)
library(imager)
```

We are using "magick" only for montage and "imager" for extracting brightness because it is easier to work with for that purpose.

12.6 Step 2: Generate Brightness Values

Within a loop that iterates through the rows, we load each respective file into the "img" object if a thumbnail exists. Then, we calculate the mean brightness and write it to a new column "$thumb_mean."

```
# Ensure the column exists and is initialized with NA
Artworks_Foto_ee$thumb_mean <- NA

# Process only rows where thumb_OK is TRUE
valid_rows <- which(Artworks_Foto_ee$thumb_OK == TRUE)

# Loop through the valid rows and calculate the mean brightness
for (i in valid_rows) {
  img <- load.image(Artworks_Foto_ee$thumb_name[i])
  Artworks_Foto_ee$thumb_mean[i] <- mean(img)
}
```

This only needs to be done once. Now we have the brightness values in the data frame, ready for sorting.

12.7 Step 3: Create and Sort the File List

Next, we create a file list containing only the thumbnails that exist and sort them by "$thumb_mean." The "subset" function in this context is used to filter "Artworks_Foto_ee," including only rows where the "thumb_OK" column is "TRUE" (see the chapter on subsets above).

```
file_list <- subset(Artworks_Foto_ee, Artworks_Foto_ee$thumb_OK)
file_list <- file_list[order(file_list$thumb_mean, decreasing
= TRUE),]
```

Alternatively, you can sort by any other criteria, such as creation date:
file_list <- file_list[order(file_list$Date, decreasing = TRUE),]

ℹ The dates are provided in various formats, so when sorting, date formats that start with a character (e.g., "Summer 1965," "c. 1967") are listed first, without considering the actual year.

12.8 Step 4: Load and Montage Images

Now, we load these images from the directory:

```
image_list <- image_read(file_list$thumb_name)
```

After that, we create a montage of this image list and a montage of that montage (Fig. 12.2):

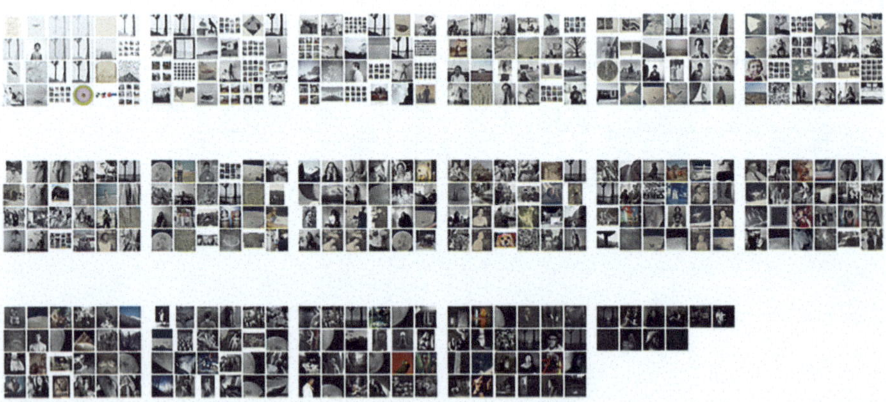

Fig. 12.2 A montage of images sorted by mean brightness value

```
image_montage(image_montage(image_list))
stable_montage <- image_montage(image_list, tile = "x34",
geometry = "100x100")
image_write(stable_montage, "stable_montage.jpg", format = "jpg")
```

This code creates a montage from the list of images, arranging them into 34 rows (using "tile = 'x34'") with each image sized to "100x100" pixels ("geometry = '100x100'"). The resulting montage (Fig. 12.3) is then saved as a JPEG file. Feel free to customize this script according to your interests. You can work not only with square photos, but also with a subset of images that you find relevant.

> If you are interested in extracting color values from images further, look at the script of the Cleveland Fun Facts Bot for more inspiration.

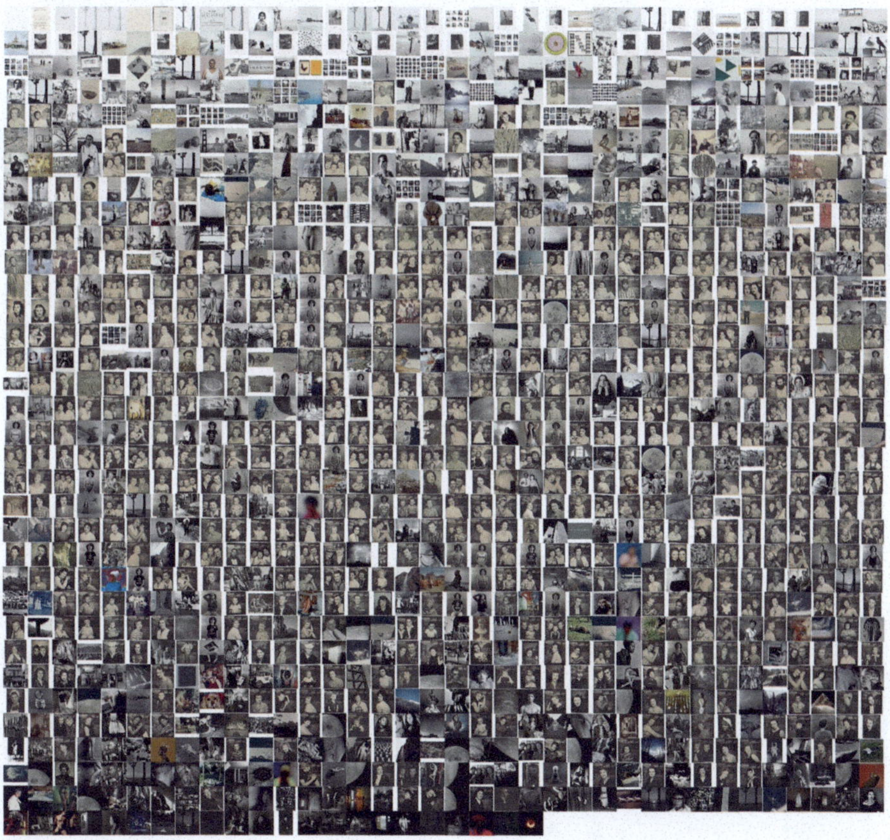

Fig. 12.3 A grid of images arranged by mean brightness

Chapter 13
Case Study 4: Image Plot

In the context of cultural data sets, an image plot is a visual representation that uses images as data points on a plot to convey insights, patterns, or relationships within the dataset. It is a way to visually explore and analyze cultural data by positioning images based on specific attributes or criteria associated with them.

To create such a two-dimensional plot, we process the overall brightness of an image and calculate that by averaging the pixel values' intensity. This single value provides a quick and straightforward way to assess an image's brightness. Images with higher average brightness will appear lighter, while those with lower values will appear darker. Another set of features are the RGB values, short for red, green, and blue values. They represent the color of each pixel in an image. They are used for analyzing and manipulating color information in images. A third set of features, easy to derive from an image, are the HSV values that represent Hue, Saturation, and Value in an image:

- Hue (H) describes the dominant wavelength of color (from 0 to 360° on a color wheel).
- Saturation (S) is the intensity or purity of color (0–100%).
- Value (V) indicates the amount of light in the color (0–100%).

If you have downloaded images via URL and want to extract RGB values and brightness (average), look at a code snippet (part of a Tweetbot project). To convert RGB (Red, Green, and Blue) to HSV (Hue, Saturation, Value), you can use the "rgb2hsv" function.

Once you have downloaded the images into a subdirectory (e.g., "imgs_moma"), you can extract RGB and HSV data from them and position them on a plot based on these values and the creation year. Downloading might take some time, but you can interrupt the download process and work with only a subset of the images for testing purposes. This allows you to test the extraction and visualization steps without waiting for all images to be downloaded.

H. Klinke, *Cultural Data Science*, Quantitative Methods in the Humanities and Social Sciences, https://doi.org/10.1007/978-3-031-88130-5_13

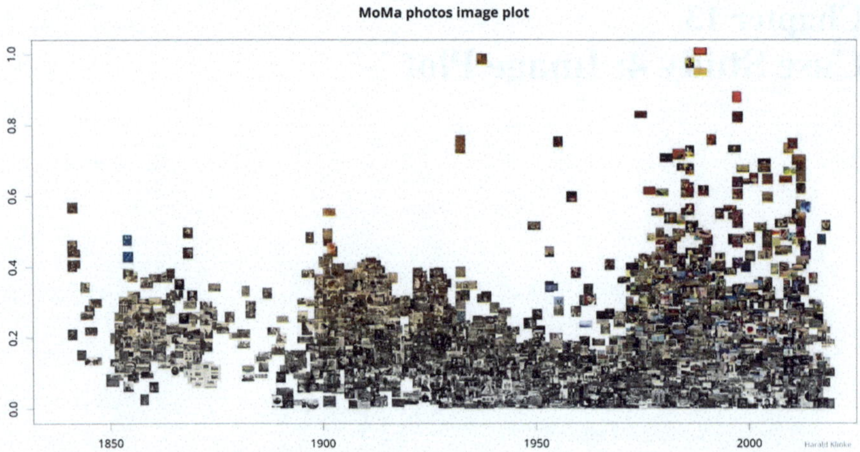

Fig. 13.1 Image plot showing the saturation of photos over time

In the above image (Fig. 13.1), you see a subset of photographs plotted by color saturation and creation year. Early photography probably used sepia-toned paper, there is a collection gap at the end of the nineteenth century (curators, take note!), around 1900 black-and-white photography on color-stable paper emerged, and from 1950 onward, there are increasingly high-saturation images while black-and-white photography continued.

If you have not already added the required additional columns, you can do so as follows:

```
Artworks_photo <- subset(Artworks, Artworks$Department ==
"Photography")
loop_length <- nrow(Artworks_photo)

for (i in (1:loop_length)) {
  filename <- paste("imgs_moma\\", Artworks_photo$ObjectID[i],
".jpg", sep = "")
  if (file.exists(filename)) {
    image <- load.image(filename)
    dimensions <- dim(image)
    Artworks_photo$img_height[i] <- dimensions[1]
    Artworks_photo$img_width[i] <- dimensions[2]
    Artworks_photo$img_avg_brig[i] <- mean(image)

    Artworks_photo$img_red[i] <- mean(R(image))
    Artworks_photo$img_green[i] <- mean(G(image))
    Artworks_photo$img_blue[i] <- mean(B(image))
```

```
    img_hsv <- rgb2hsv(
      Artworks_photo$img_red[i],
      Artworks_photo$img_green[i],
      Artworks_photo$img_blue[i]
    )

    Artworks_photo$img_h[i] <- img_hsv[1]
    Artworks_photo$img_s[i] <- img_hsv[2]
    Artworks_photo$img_v[i] <- img_hsv[3]

    message(round(i/loop_length*100, digits = 0), "%"," ",i,"
",filename )
  } else {
    message(i, " ","no image")
  }
}
```

Explanation: First, we create a subset if desired. Then, we loop through it. We generate the filename from the Object-ID in each iteration. If the file exists, we load the respective image from the hard drive. We extract the dimensions, RGB values, and HSL values and write them into additional rows (you may need to create the columns as empty before the loop with Artworks_photo$img_height = NA, etc.).

And here is the code for the image plot:

```
library(imager)
png(
  filename = "imageplot_MoMa_photos.png",
  width = 1920, height = 1080, pointsize = 24,
  bg = "white"
)
plot(c(1839, 2018), c(0, 1), type = "n",
     xlab = "", ylab = "saturation", main = "MoMa photos
image plot")

for (i in (1:loop_length)) {
  filename <- paste("imgs_moma\\", Artworks_photo$ObjectID[i],
".jpg", sep = "")
  if (file.exists(filename) &
      !is.na(Artworks_photo$cdate[i]) &
      !is.na(Artworks_photo$img_s[i])) {
    image <- load.image(filename)
    year <- Artworks_photo$cdate[i]
    value <- Artworks_photo$img_s[i]
```

```
    x_plus <- Artworks_photo$img_height[i] / 100
    y_plus <- Artworks_photo$img_width[i] / 10000
    rasterImage(image, year, value, year + x_plus, value +
y_plus)
  } else  {
    message("no data")
  }
}
dev.off()
```

And here is the explanation:

We start by writing directly to a PNG file, which is faster in terms of performance. Then, we create an empty plot from the founding year to the present. In a loop, we go through the table of the Artworks_photo subset (`loop_length` is defined earlier as the number of rows in the set). We define the path to the file, which is located in the "imgs_moma" subdirectory (adjust this path according to your directory structure). Only if this file exists on the hard drive, and we have a creation date and a saturation value (which we previously wrote to the img_s column), we load the image, extract the creation year and saturation value. Otherwise, we write "no data." `x_plus` and `y_plus` are scaling factors used to place the images on the plot without displaying them in full size. Using `rasterImage`, we position the image based on the year and saturation value. Once the loop is complete, we close the image file with dev.off().

And these are the Cleveland photos (only up to 1925) plotted by Hue. Sepia is particularly evident (Fig. 13.2).

> Another way to place images on a plot is using the library called ggimage.

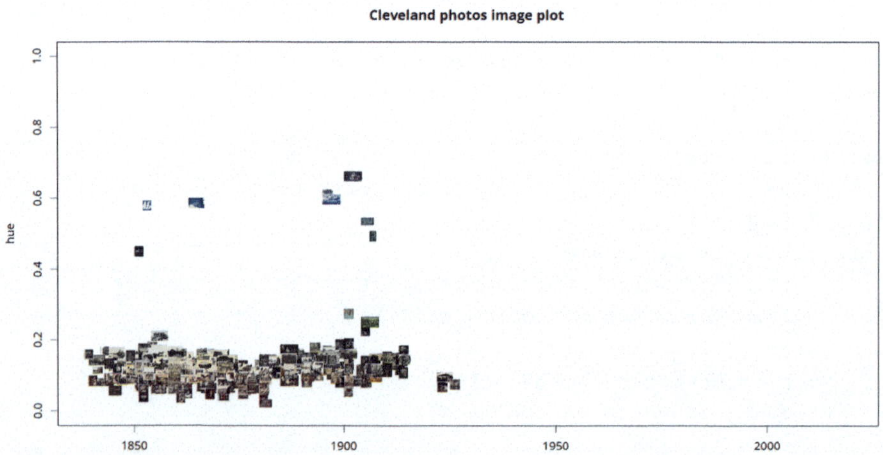

Fig. 13.2 Image plot showing the hue of photos in the Cleveland Museum of Art collection

Chapter 14
Dimensions

In plotting and data analysis, the term "dimensions" often refer to the attributes or variables that represent data points. They play a key role in data representation, visualization, analysis, and feature engineering. Dimensions help create plots, segment data, build models, and reveal patterns within datasets.

When you have a list of numerical values, you can create a histogram to visualize the distribution of those values. Histograms are useful for understanding the frequency and spread of data points within a single variable.

When you have two lists of values, you can create an XY plot, also known as a scatter plot. This type of plot allows you to explore the relationship between two variables, which can help identify correlations or patterns. For example, this could be the length and width of an image.

With three dimensions or three sets of data, you can venture into 3D visualization, which adds depth to your plots. 3D plots can be used to visualize how three variables interact with each other.

Alternatively, you can use time itself as a visualization tool, creating dynamic plots that show how data changes over time. Time-series data visualizations are valuable for tracking trends, patterns, and anomalies over a period.

> For further inspiration and guidance, you can watch the videos by Hans Rosling to explore various data visualization techniques and get ideas for effectively presenting your data.

H. Klinke, *Cultural Data Science*, Quantitative Methods in the Humanities and Social Sciences, https://doi.org/10.1007/978-3-031-88130-5_14

Here is an example related to the size of photographs in MoMA: Fig. 14.1 indicates that smaller media typically represent gelatin silver prints, while larger ones are often C-prints.

The visualization below (Fig. 14.2) shows the evolution of photograph dimensions in MoMA over time. It becomes apparent that until the mid-20th century, the dominant aspect ratio was 4:3, later transitioning to 3:2 for gelatin silver prints.

?What other insights can be gleaned from this data?

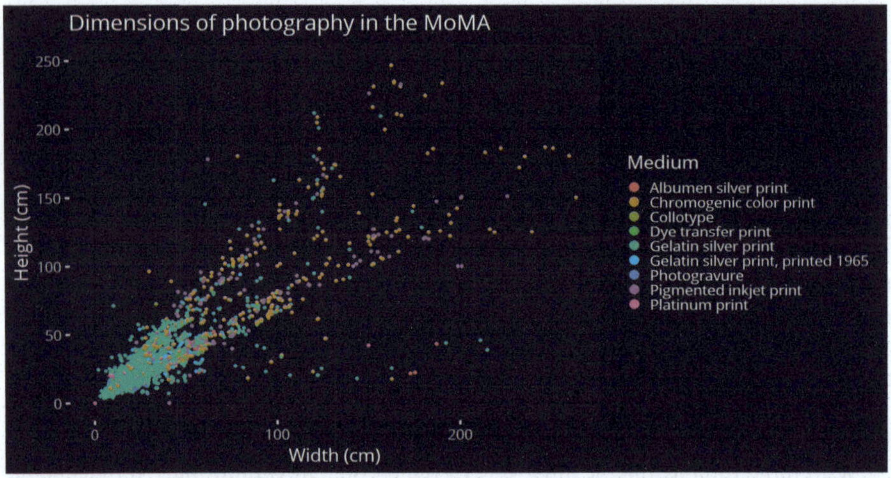

Fig. 14.1 Scatter plot showing the dimensions of photography in the MoMA collection

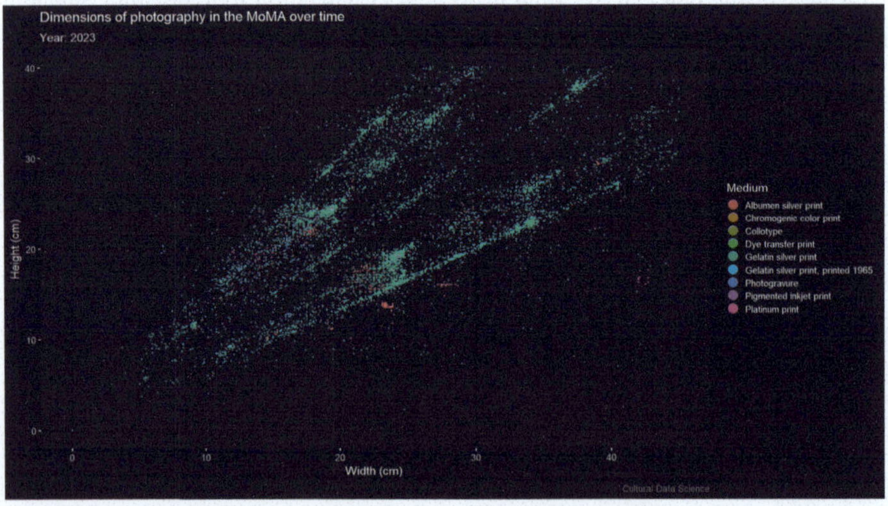

Fig. 14.2 Scatter plot illustrating the dimensions of photography over time

This visualization was created using the gganimate package. Here is the code:

```
library(ggplot2)
library(gganimate)
library(ggdark)

# List of relevant media
relevant_media <- c("Albumen silver print",
                    "Chromogenic color print",
                    "Collotype",
                    "Dye transfer print",
                    "Gelatin silver print",
                    "Gelatin silver print, printed 1965",
                    "Photogravure",
                    "Pigmented inkjet print",
                    "Platinum print")

# Subset the Artworks dataset
photography <- subset(Artworks, Medium %in% relevant_media)

# Plot
p <- ggplot(photography, aes(x = Width..cm., y = Height..cm.,
col=Medium , fill = Medium)) +
  xlim(0, 45) +
  ylim(0, 40) +
  geom_point(size=4, alpha = .85) +
  transition_states(as.integer(cdate),
                    transition_length = 1,
                    state_length = 2, wrap = F) +
  transition_time(as.integer(cdate)) +
  enter_fade() +
  exit_fade() +
  labs(
    title = "Dimensions of photography in the MoMA over time",
    subtitle = "Year: {frame_time}",
    x = "Width (cm)",
    y = "Height (cm)",
    caption = "Cultural Data Science by H. Klinke" ) +
  shadow_mark(alpha = 0.5, size = 1.5) +
  dark_theme_gray(base_family = "Arial", base_size = 25 ) +
  theme
(plot.caption=element_text(color=rgb(1,1,1,.4)) ) +
  guides(colour = guide_legend(override.aes = list(size=10)) )
```

```
# Render the animation and store it in a variable
animation <- animate(
  p,
  renderer = ffmpeg_renderer(),
  duration = 30,
  fps = 20,
  width = 1920,
  height = 1080
)

# Specify the output directory and file name
output_file <- "animation.mpg"

# Save the animation
anim_save(output_file, animation = animation)
```

> ⓘ Make sure that ffmpeg is properly installed and available on your system, as it is required to render the video file.[1] For Mac users: Install Homebrew (if not already installed). This is a package manager for macOS that simplifies the installation of software like ffmpeg.

This example displays how data visualization can reveal patterns and trends in photography dimensions over time, providing valuable insights for art historians and researchers at MoMA.

> ⓘ Note: You can comment out a line of code by adding a "#" symbol at the beginning. To comment out multiple lines, select them and press Ctrl+Shift+C. For code formatting, you can use Ctrl+A to select all, and then Ctrl+Shift+A to format it neatly. Both functions can also be found in RStudio in the menu under "Code."

Dimensionality reduction in the context of features from images, such as t-SNE (t-distributed stochastic neighbor embedding), is a technique used to reduce the number of dimensions or variables in a dataset while preserving its essential structure and relationships. It is commonly employed in machine learning, data analysis, and image processing to visualize high-dimensional data or to prepare it for downstream tasks like clustering or classification.

[1] https://www.ffmpeg.org/download.html.

t-SNE is a specific dimensionality reduction algorithm that excels at preserving the local and global structure of data points when mapping them from a high-dimensional space to a lower-dimensional space (usually 2D or 3D).

We do not go into details here how this works mathematically, but it will become apparent that by reducing the dimensionality of high-dimensional image feature data using techniques like t-SNE, it is easier to visualize and analyze complex datasets, identify patterns, and gain insights that might not come to our attention in the original high-dimensional space. It is particularly useful for image analysis, where images are represented by numerous features, making it challenging to explore and understand the data in its original form.

In R, you can accomplish dimensionality reduction and create t-SNE image plots for images and their derived features using several libraries and packages. Here is a step-by-step guide on how to do this:

First, you can simply use numerical features from the metadata, or, for a more advanced method, you can load and preprocess your image data, by importing the image data into R using packages like "jpeg," "png," or "imager." Extract relevant features from the images, such as color histograms, texture features, or deep learning-based features using packages like "imager," "EBImage," or "keras." The extracted features for every object should be stored in a matrix named your_feature_data for further analysis.

Then we perform t-SNE Dimensionality Reduction. To do so, install and load the "Rtsne" package, which provides an R interface to the Barnes-Hut implementation of t-SNE. Apply the t-SNE algorithm to your feature data to reduce its dimensionality. For example:

```
library(Rtsne)
tsne_result <- Rtsne(your_feature_data, dims = 2)
```

Now we have reduced the number of dimensions to two. Now we can create t-SNE image plots by using the "ggplot2" package to create scatterplots for visualizing the t-SNE results (Fig. 14.3).

> ℹ If you like you can combine the t-SNE results with other metadata, if available, to create informative plots.

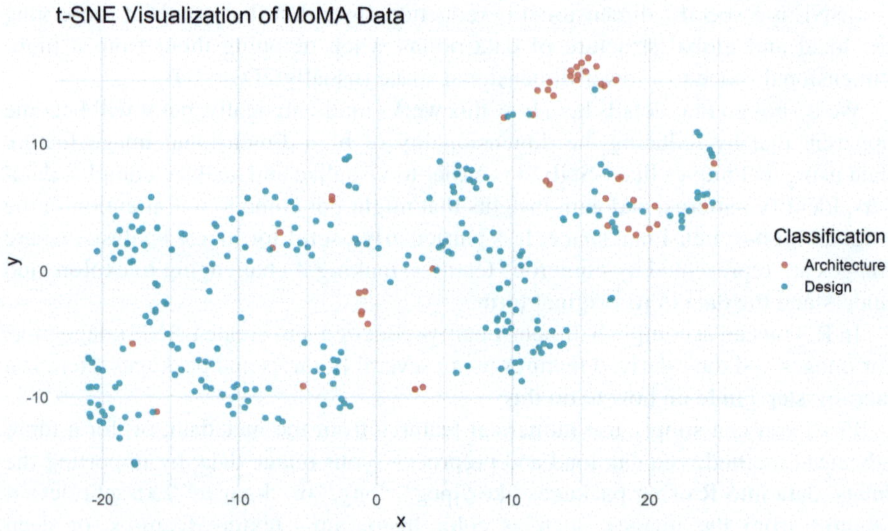

Fig. 14.3 t-SNE visualization of MoMA data, showing a clear distinction between Architecture (red) and Design (blue) categories

Here is the complete code:

```
library(Rtsne)
library(ggplot2)
library(ggimage)

# Step 1: Subset the first 2000 rows from Artworks
artworks_subset <- Artworks[1:2000, ]

# Step 2: Identify numerical columns for t-SNE
numerical_columns <- c("adate",
       "cdate",
       "AspectRatio",
       "Width..cm.",
       "Height..cm.",
       "Depth..cm.")
numerical_data <- artworks_subset[, numerical_columns]

# Step 3: Remove rows with NA values
numerical_data <- na.omit(numerical_data)

# Step 4: Convert to matrix for t-SNE
feature_data <- as.matrix(numerical_data)

# Step 5: Keep the Classification column for coloring
artworks_clean <- artworks_subset[rownames(numerical_data), ]
categories <- artworks_clean$Classification
```

```
# Step 6: Perform t-SNE
# Remove duplicate rows from feature_data
feature_data <- unique(feature_data)

# Optional: Adjust the associated rows in artworks_clean and
categories
artworks_clean <- artworks_clean[rownames(feature_data), ]
categories <- artworks_clean$Classification

tsne_result <- Rtsne(feature_data, dims = 2, perplexity = 20)


# Step 7: Combine t-SNE results with categories
tsne_df <- data.frame(x = tsne_result$Y[, 1],
      y = tsne_result$Y[, 2],
      Classification = categories)

# Step 8: Visualize t-SNE colored by classification
ggplot(tsne_df, aes(x = x, y = y, color = Classification)) +
  geom_point(size = 2, alpha = 0.8) +
  labs(title = "t-SNE Visualization of MoMA Data", color =
"Classification") +
  theme_minimal(base_size = 15)+
  theme
(
    panel.background = element_rect(fill = "white", color = NA),
# White plot panel
    plot.background = element_rect(fill = "white", color = NA)
# White entire plot area
  )

ggsave("tsne_plot.png",
       plot = last_plot(),
       width = 10,
       height = 6)
```

Mind that in t-SNE visualizations, the coordinates x and y do not represent actual measurements or scales; they are only abstract values that indicate relative similarities or differences between data points in a reduced-dimensional space.

Perplexity is a parameter that controls the balance between local and global data structure. A low perplexity (e.g., 5) focuses on local structure (small clusters or neighborhoods). A high perplexity (e.g., 50) captures more global relationships across the dataset. The value should be smaller than the number of data points, as it directly relates to the effective neighborhood size.

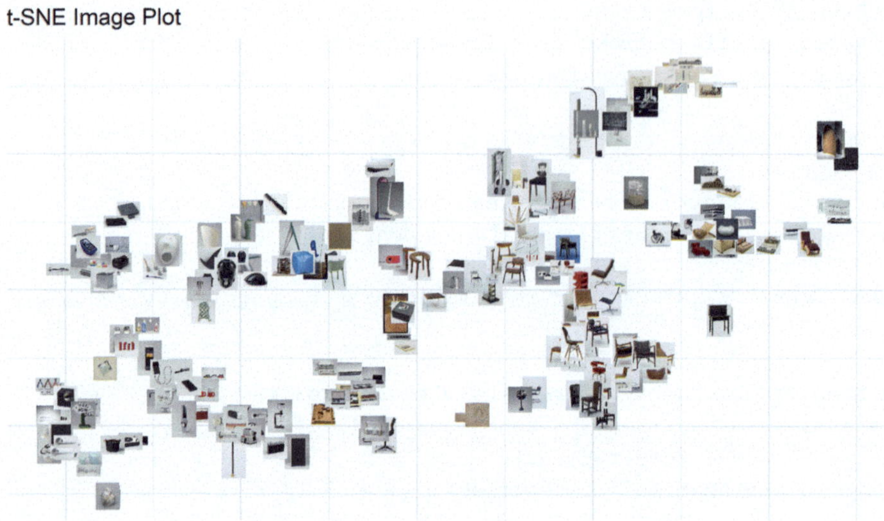

Fig. 14.4 t-SNE image plot showcasing a visual representation of MoMA's collection, where similar items are grouped closely together based on their features

Finally, we plot our images in this t-SNE space (Fig. 14.4). Overlay or annotate the t-SNE plot with images to visualize their positions in the lower-dimensional space. You can use the "ggimage" package for this.

```
# Step 9: Add thumbnail paths for images
image_dir <- " imgs_moma"

# Construct the file paths using ObjectID
image_paths <- file.path(image_dir, paste0(artworks_
clean$ObjectID, ".jpg"))

# Ensure valid paths
valid_indices <- file.exists(image_paths)
image_paths <- image_paths[valid_indices]
tsne_df <- tsne_df[valid_indices, ]
tsne_df$image_path <- image_paths

# Step 10: Create t-SNE plot with overlaid images
tsne_plot <-  ggplot(tsne_df, aes(x = x, y = y)) +
  geom_image(aes(image = image_path), size = 0.05) +  # Adjust
size for thumbnails
  labs(title = "t-SNE Image Plot", x = NULL, y = NULL) +
  theme_minimal(base_size = 15)+
```

```
  theme
(
    panel.background = element_rect(fill = "white", color = NA),
# White plot panel
    plot.background = element_rect(fill = "white", color = NA) ,
# White entire plot area
    axis.text = element_blank()
       #      axis.title = element_blank()  # Remove axis titles

  )


# Save the plot
ggsave(
  "tsne_image_plot.png",
  plot = tsne_plot,
  width = 10,
  height = 6
)
```

Objects that are similar based on their features are grouped into clusters. This indicates that objects with similar dimensions, proportions, or other features share similar positions in the reduced-dimensional space. The plot demonstrates the variety in the collection, ranging from furniture and household items to artworks and design objects. Different object types form distinct but occasionally overlapping clusters, reflecting shared or unique feature combinations. Densely packed areas indicate many objects with very similar features. Sparse regions might represent unique or outlier objects with less common attributes. Clusters could hint at subcategories within the collection, such as chairs grouped by style, or kitchen utensils grouped by functionality. Isolated points or small groups may represent unique items that do not fit into the larger trends or categories.

If you like, you can now customize and export the resulting plot. Customize your t-SNE image plot by adjusting labels, colors, and other visual elements as needed. Analyze the t-SNE image plot to identify clusters, patterns, or relationships among your images and features. Use the plot to gain insights into your dataset, detect outliers, or guide further analysis. This process allows you to visualize complex image data in a lower-dimensional space, making it more interpretable and suitable for further analysis.

> i Besides t-sne, there are many means to reduce dimensions for visualizing. Among them, principal component analysis (PCA) and uniform manifold approximation and projection (UMAP) are two other common dimensionality reduction techniques. All have their strengths, and the choice between them depends on the characteristics of your data and the goals of your analysis.

Chapter 15
Network Visualizations

Representing data as a network can help illustrate the centrality of the Stern Collection or the sources of Picasso's works. I used the Gephi program for this purpose. However, in R, you can achieve the same results using packages like igraph (examples, details), qgraph, or network.

In essence, networks consist of objects (nodes) and the connections between them (edges). Both can be extracted from MoMA's data, such as artworks or their artists and acquisition sources. Additionally, you can calculate weights, like the number of artworks coming from a source. This information can be used to determine the thickness of lines, the size of labels, or the color of nodes. Network indicators, such as centrality (how important a source or artist is due to the number of connections), can also be calculated.

Figures 15.1, 15.2, and 15.3 exemplify the epistemic potential of network visualizations by offering deeper insights into the interconnections between MoMA's artworks and their sources. By mapping these relationships, the visualizations allow us to see how artworks, artists, donors, and acquisition sources converge, revealing patterns that can otherwise be overlooked in traditional data representations.

The first image highlights Pablo Picasso and the various sources contributing to his collection at MoMA. By representing the donors, funds, and specific collections associated with Picasso's works, this network plot not only visually identifies the central role Picasso plays in the museum's collection but also reveals the pathways through which his art was acquired. This type of analysis can be used to understand how an artist's influence is spread across multiple acquisition sources, with nodes and edges symbolizing the intricate web of connections that brought Picasso's works into the museum.

The second image shifts focus to "The Louis E. Stern Collection," illustrating the relationships between artists such as Picasso, Henri Matisse, Joan Miró, and Georges Braque. The central node, representing the collection, radiates outward to these artists, effectively mapping how this collection brings together influential figures and their works. The use of centrality measures in this case could highlight which artists have the most substantial impact within the collection.

H. Klinke, *Cultural Data Science*, Quantitative Methods in the Humanities and Social Sciences, https://doi.org/10.1007/978-3-031-88130-5_15

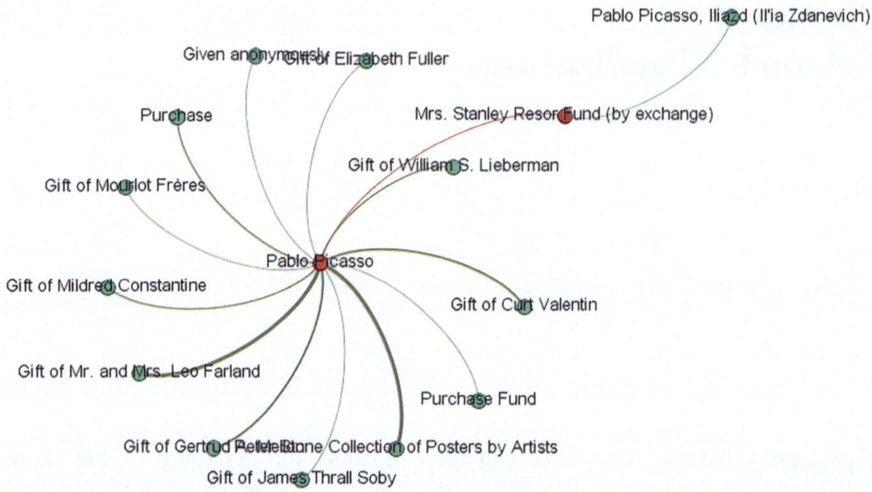

Fig. 15.1 Network plot showing Pablo Picasso and its related artworks

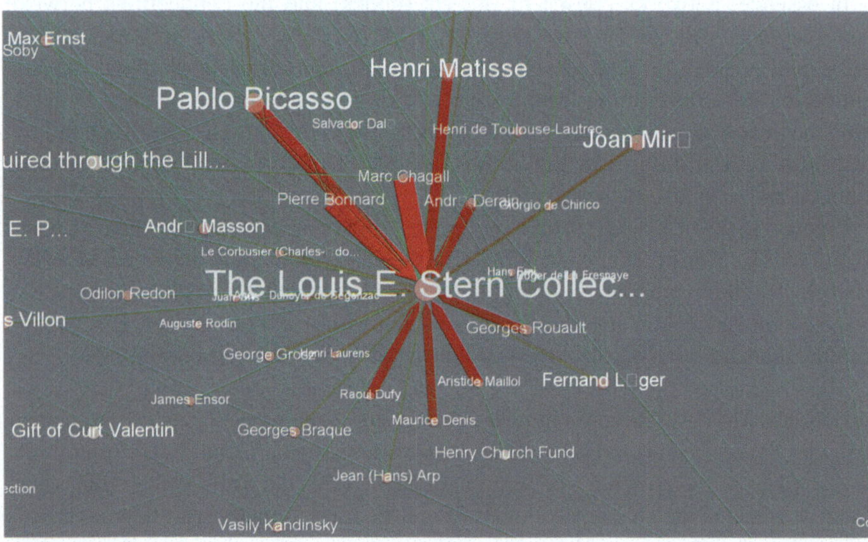

Fig. 15.2 Network visualization of the Louis E. Stern Collection

The third visualization further expands the scope by representing connections between various art acquisitions, collections, and donors. It features artists like Ellsworth Kelly and Willem de Kooning, with nodes signifying contributors or acquisitions, while edges illustrate the relationships that led artworks to the museum. The analysis of such networks can reveal which sources and individuals play a more significant role in shaping the museum's collection trajectory, as well as how acquisitions relate to the museum's strategic growth over time.

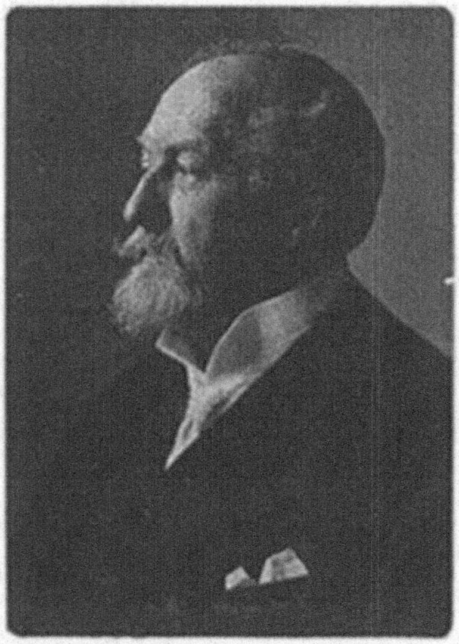

Fig. 15.3 Portrait of Otto Wagner, an Austrian architect and urban planner (public domain image from Wikipedia)

15.1 Linked Open Data: Wikidata and ULAN

How can you identify artists without always having to spell their names correctly? You can use an ID, a unique number that identifies the artist. Fortunately, we have such IDs included in MoMA's data.

In MoMA's data, you will find references to ULAN IDs and Wikidata QIDs. These are links to data on Getty Institute and Wikimedia servers:

- ULAN IDs refer to the Getty Vocabulary Program's Union List of Artist Names (ULAN), a structured database of artist names and related information maintained by the Getty Research Institute. ULAN IDs provide a standardized way to reference artists across collections and research.
- Wikidata QIDs are unique identifiers used in Wikidata, a free and collaborative knowledge base maintained by the Wikimedia Foundation. Each QID corresponds to a specific entity like a person, artwork, or concept, enabling connections to structured data about that entity.

Sometimes, a list of artists may not exactly match the naming conventions of a particular collection. How can this be resolved?

- By manual spell checking.
- Or by assigning internationally unique IDs to each artist.

For the latter, there are Authority Files, institutions that assign unique IDs. Thus, you compare only the IDs and not the names. If all museums followed this practice, it would make things easier for data scientists.

For example, in MoMA's data (Artists.csv), you can find references to both Authority Files, QID and Wikidata.[1]

[1] https://github.com/MuseumofModernArt/collection/blob/main/Artists.csv.

15.2 Locating the ID

Returning to our MoMA data from "artworks," the first entry is "Ferdinandsbrücke." It contains an entry for the artwork (URL). It also contains an entry for the artist's name, which is "Otto Wagner." The next column is "ConstituentID," which is "6210." This is the artist's ID. The "artists" data also contains links to artists' names via these "ConstituentID" values. If you search for "6210" there, you will get "Otto Wagner."

Now, let us look at the last columns in "artists" and find the following:

- ULAN: Look up 500016971, and you will get "Otto Wagner."
- QID: Take Q84287, and you will get "Otto Wagner" again, but this time with a portrait of him.

15.3 Explanation

Unlike Wikipedia, Wikidata consists of structured data that can be queried automatically. In contrast to our MoMA dataset, which we download, Wikidata has a server interface that can be queried. To see what Wikidata is, you can experiment with the Wikidata Query Service.[2] For example, under "Filter," type "creator (P170)" and "Albrecht Dürer (Q5580)" as the value (this is the QID!).

This will give you a query for all works in Wikidata for which Dürer is the creator (Fig. 15.4). Click the triangle to execute it.

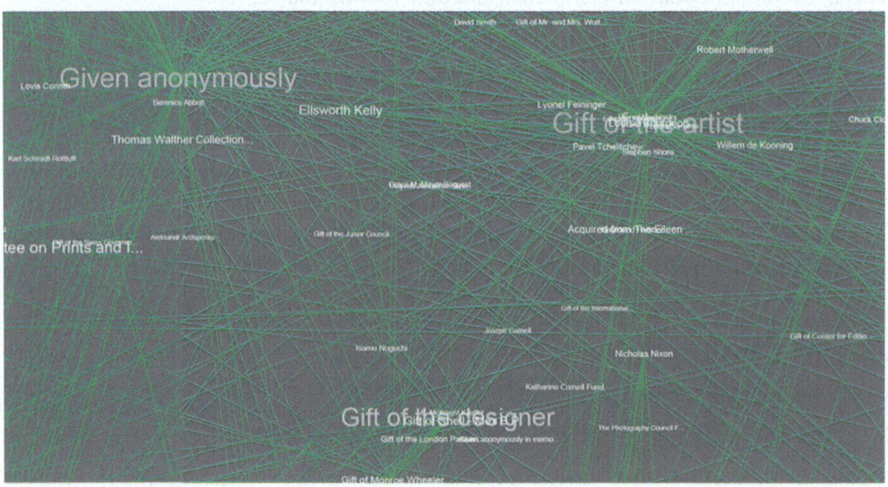

[2] https://query.wikidata.org/ or alternatively https://query-main.wikidata.org/sparql.

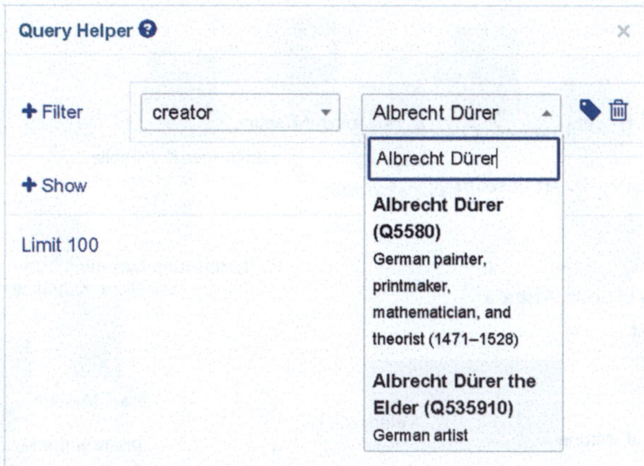

Fig. 15.4 Search query in the Wikidata Query Helper

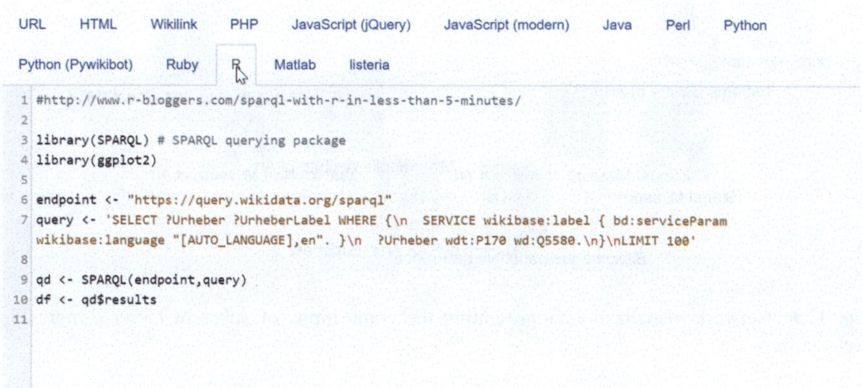

Fig. 15.5 Code snippet to execute a SPARQL query against Wikidata

Look at this list and try using a QID from our MoMA data.

Now, we do not want to do this through the web interface but with R. It is as simple as clicking "Code" on the right and then "R" (Fig. 15.5).

You get the R code that you can simply copy and paste into a script. Done! (You might need to install the SPARQL and ggplot2 packages beforehand, but you know how to do that):

```
install.packages("SPARQL")
install.packages("ggplot2")
```

The "Endpoint" is the server I want to query, and "Query" is my query. The results could be written into a new table.

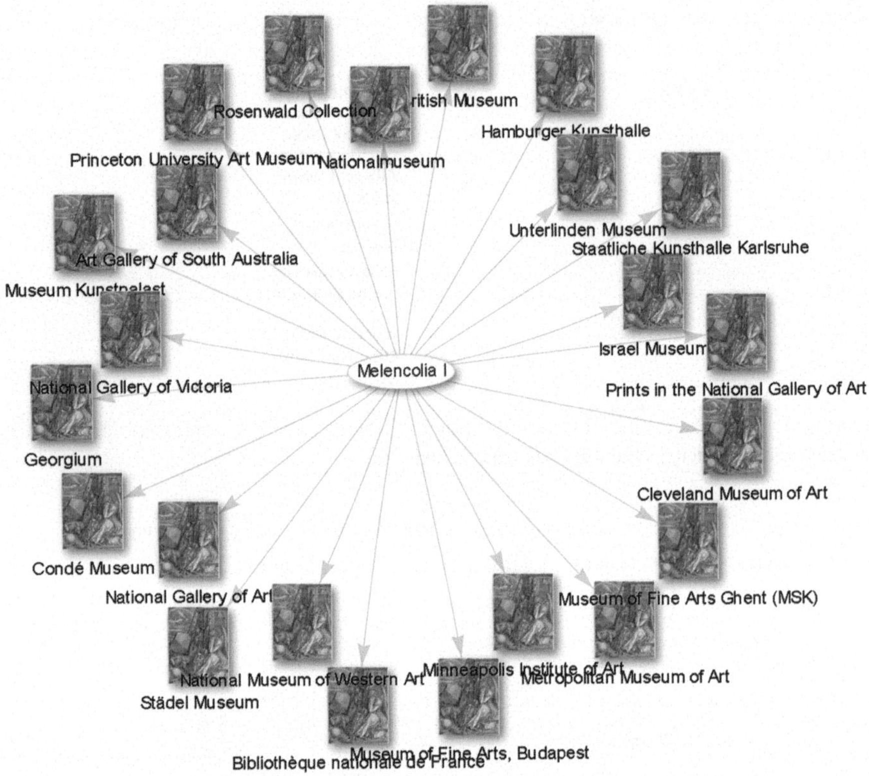

Fig. 15.6 Network visualization, representing the connections of Albrecht Dürer's engraving Melencolia I

Using ULAN works similarly, but you need to construct the query yourself. However, you can look here for more information and click on "Albrecht Dürer" to see what information you could gather about the artist there. (The presentation of the student relationships you find in it would probably resemble the network topology of a directed tree.)

This way, we can fetch additional information and expand beyond the scope of our MoMA data.

Try searching for the work Q1362177 on Wikidata to retrieve its associated collections and images. Wikidata will display all collections that house "Melencolia I" (Fig. 15.6). By selecting "Graph" on the left panel, you can visualize the results. Of course, you can query these collections and images also via code using R to interact with the Wikidata SPARQL endpoint.

Chapter 16
Image Analysis

In the previous chapters, we primarily worked with metadata in a data frame. In addition, we have also turned our attention to images to generate an Image plot using features like average brightness. But image analysis can be even more comprehensive, and it can be implemented using R libraries or cloud services (especially those requiring more computational power).

How about:

- Detecting dominant colors (how does color composition change?)
- Labeling (what themes do images have?)
- Face recognition (how many people are in the image?)
- Object recognition (is there a cat in the image?)
- Text recognition (what is written in the image?)

All this falls under the category of Visual Computing or Image Recognition.

16.1 Dominant Colors

Let us consider the following image from the Cleveland collection (Fig. 16.1):

Here, we would like to know the exact RGB values of the dominant colors. You can see how this can be done here. For this, we use the `RImagePalette` library (the path to the image is specified in `image_url`).

H. Klinke, *Cultural Data Science*, Quantitative Methods in the Humanities and Social Sciences, https://doi.org/10.1007/978-3-031-88130-5_16

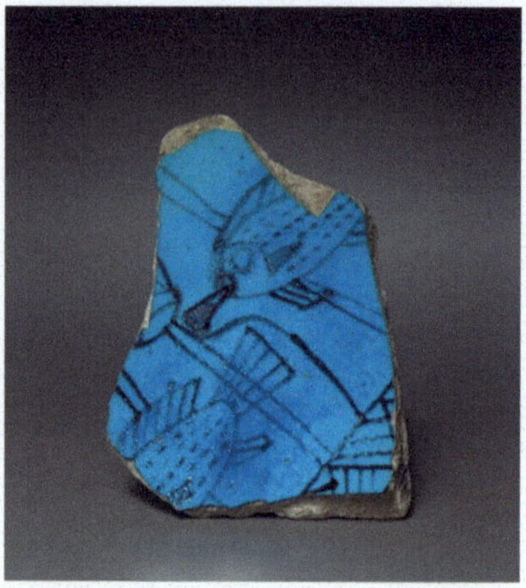

Fig. 16.1 Marsh Bowl Fragment, 1479–1429 BCE (Cleveland Museum of Art)

First, we read the respective image, create a palette with four colors, and display it:

```
# Load necessary libraries
library(httr)
library(jpeg)
library(RImagePalette)

# Define the image URL
image_url <- data_cleveland$image_web[1238]

# Read image from the web
response <- GET(image_url)
if (http_error(response)) {
  stop("Failed to download the image from the URL.")
}

# Convert raw content to image format
img <- readJPEG(content(response, as = "raw"))

# Extract color palette
imgPalette <- image_palette(img, n = 4)

# Display colors
scales::show_col(imgPalette)
```

The result will show you the dominant colors (Fig. 16.2):

The RGB hex values from "imgPalette" can be added as additional columns to your data frame, giving you derived features that can be used for analysis.

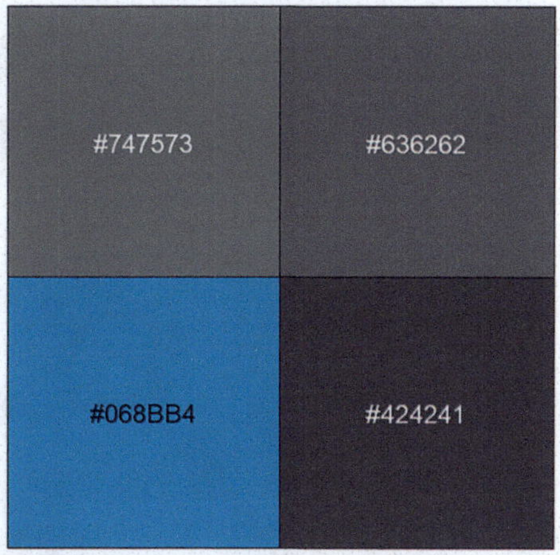

Fig. 16.2 A color palette featuring dominant colors

16.2 Face Recognition

Now, you might wonder how many faces are in an image and perhaps even want to extract them.

In R, you can use the "image.libfacedetection" library for face detection, and ImageMagick helps with image processing.

To do this, you install and integrate the libraries if necessary, load the image (here we take another image from the MoMA dataset), apply face detection, and you will find the number of detected faces and their bounding boxes (Fig. 16.3):

```
image_url <- data_cleveland$image_web[3675]
# Read the image directly from the web
img <- image_read(image_url)

# Detect faces in the image
faces <- image_detect_faces(img)

# Print face detection result
print(faces)
```

This results in five faces:

```
$nr
[1] 5
```

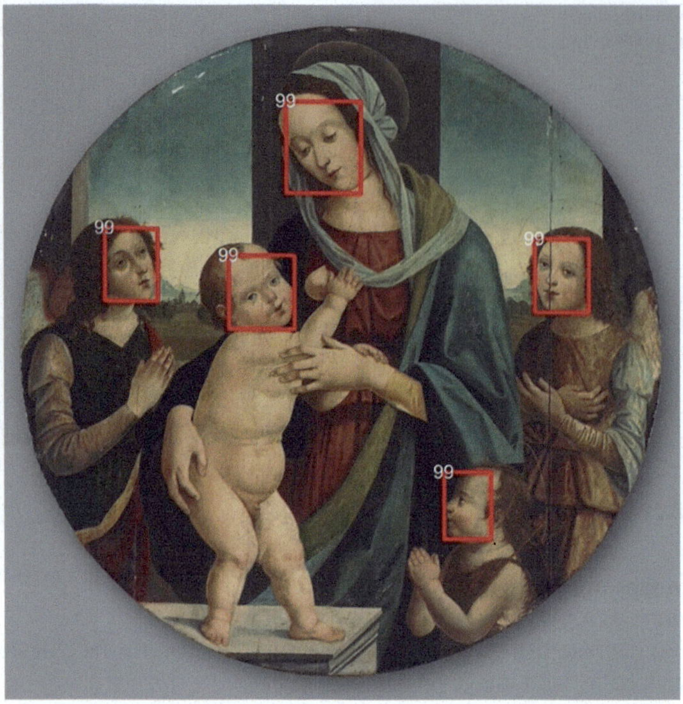

Fig. 16.3 Virgin and Child with John the Baptist and Angels, c. 1500 by Pietro del Donzello (Cleveland Museum of Art), with bounding boxes indicating the recognized face along with the confidence score

```
$detections
    x   y width height confidence landmark1_x landmark1_y
landmark2_x landmark2_y landmark3_x
1 287 322    82     95         99         319         370
344         354         349
2 359 124    96    118         99         380         184
416         174         402
3 129 287    68     94         99         154         330
180         321         181
4 677 301    68     95         99         689         336
716         342         694
5 561 601    60     86         99         578         634
585         635         566
  landmark3_y landmark4_x landmark4_y landmark5_x landmark5_y
1         375         341         398         361         385
2         210         408         222         432         213
3         340         169         361         189         352
4         355         690         369         710         373
5         647         569         666         575         666
```

The output shows the coordinates and landmarks of the detected faces' bounding boxes. It includes the x and y positions for the top-left corner of the bounding box, along with its width and height. The confidence score indicates the high certainty of the detection. In addition, landmark coordinates represent specific facial features, like eyes, nose, and mouth, providing precise locations within the face for further analysis.

You can even plot and crop the detected faces:

```
plot(faces, img, border = "red", lwd = 7, col = "white")
face <- head(faces$detections, 1) # First face
image_crop(img, geometry_area(x = face$x, y = face$y, width =
face$width, height = face$height))
```

> As a proof of concept, I have cropped all the faces from MoMA's images. You can zoom into the view on my website: https://www.harald-klinke.de/cultural-data-science.
>
> Please note the false positives (things recognized as faces but are not) and the false negatives (faces that are not recognized as such).
>
> Also, note that this is Face Detection, i.e., finding where faces are in an image (usually done today with the efficient Viola-Jones method). Face Recognition, which is identifying which person a face belongs to, is a bit more demanding.

16.3 Text Recognition

Let us take yet another image from the Cleveland collection. What if we were to analyze all images for text and use that as a basis for analysis (Fig. 16.4)?

For this, we use the Tesseract library. Load the image and apply optical character recognition (OCR):

```
# Load necessary libraries
library(tesseract)
library(magick)

# Read the image directly from the web
image_url <- data_cleveland$image_web[6485]
img <- image_read(image_url)

# Apply OCR
text <- ocr(img)
```

Fig. 16.4 Bookplate: Elizabeth Huntting, 1903 (Cleveland Museum of Art)

The result in the "text" variable applied to Fig. 1.1 will be the recognized text:

```
[1] "ee\nHunting' Book\n"
```

Again, the "\n" character in this string represents a line break (Escape).

Not bad for text recognition, right? At the same time, the limitations of text recognition become clear. Here too, the rule is: the better the source, the better the results.

Chapter 17
Cloud Computing

You can perform object recognition on your local computer and even train your own models for it. However, it is often simpler to perform these processes in the cloud using pre-trained models.

But how do you communicate with cloud servers? It is actually quite straightforward, using well-defined interfaces that usually require just a single line of R code.

17.1 Cloud

"Cloud" means there is a computer on the internet that you can interact with using code. You can send something and receive something in return, much like a web server, except you are interacting with a dedicated service that returns structured data. This is why it is typically referred to as Software-as-a-Service (SaaS).

This interaction is done through a protocol for an interface called an application programming interface (API), specifically a Rest API, which is quite simple. Figure 17.1 illustrates the principal relationships between each element in an API call, showing how a client communicates with a database through a Rest API using methods like GET, while receiving responses usually in JSON format.

In essence, we say, "Hey, here's an image, what's in it?" and the cloud service responds with something like, "I believe it's a cat, with 95% confidence, located in the lower right corner."

H. Klinke, *Cultural Data Science*, Quantitative Methods in the Humanities and Social Sciences, https://doi.org/10.1007/978-3-031-88130-5_17

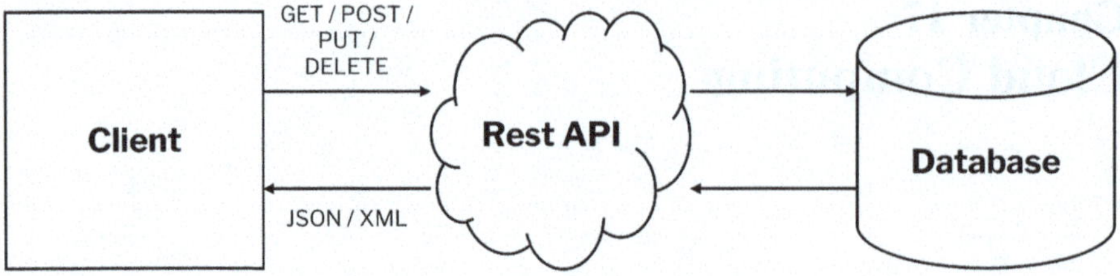

Fig. 17.1 REST API architecture

The remarkable thing is that we can automate this process. We can keep asking for answers, and for any number of images, which can serve as the basis for further analysis (e.g., how many cats are depicted in my collection?).

> Note: You see, downloading data from GitHub is just one way to handle data. Many museums offer their databases as cloud services via Rest APIs. For example, the Metropolitan Museum, the Rijksmuseum, and the Brooklyn Museum.[1]

17.2 Rest API

Let us look at an API with a simple example: The Cleveland Museum of Art offers its collection data through an API. The endpoint, which is the address where we send our queries, is `https://openaccess-api.clevelandart.org/api/artworks/`.

The types of questions you can ask the service are well documented (probably the most interesting document you will read today!): https://openaccess-api.clevelandart.org/

The simplest query is "q," which stands for query. You can search the database using keywords. For example, if you want to search for the term "uchikawa," your call would look like this:

```
https://openaccess-api.clevelandart.org/api/
artworks/?q="uchikawa"
```

Try this out in your browser!

What you receive back is data in JSON format, essentially a text file with a hierarchical structure of the provided data. We will see how to process this data shortly (Fig. 17.2).

[1] https://metmuseum.github.io/, https://data.rijksmuseum.nl/object-metadata/api/, https://www.brooklynmuseum.org/opencollection/api/docs/overview.

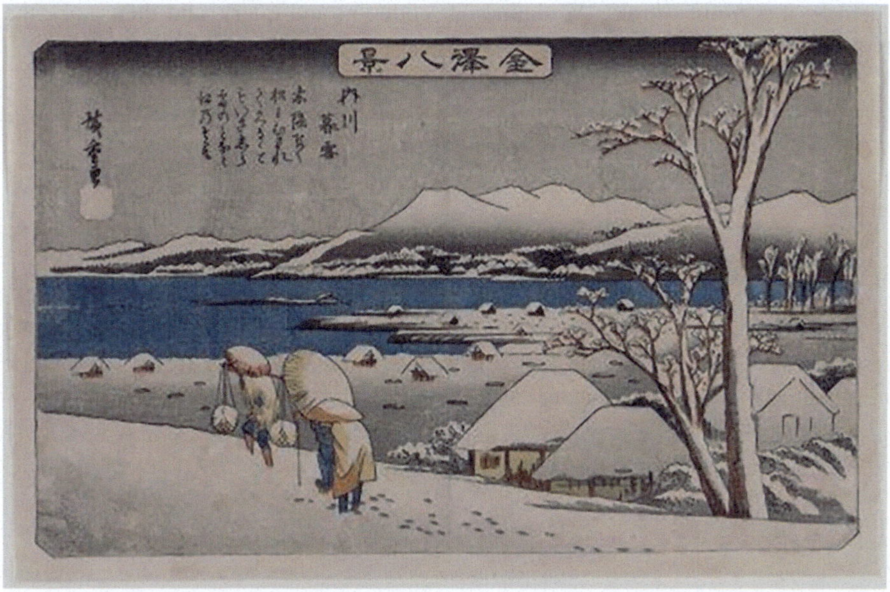

Fig. 17.2 Utagawa Hiroshige (c. 1835–1836): evening snow at Uchikawa (Cleveland Museum of Art)

What is crucial is that the structure is consistent, meaning we can expect the same information to be found at the same location every time. For instance, we can expect the title of the retrieved work to be under data/title: "`Evening Snow at Uchikawa, from the series Eight Views of Kanazawa`."

Or the path to the image under data/images/web/url: `https://openaccess--cdn.clevelandart.org/1940.987/1940.987_web.jpg`.

Take a closer look at the data structure; it is quite fascinating!

In this example, we have only one result, but often you will receive multiple results. This is not meant for human reading; it is intended for computer processing. And as you know, we can process it with R.

> Please keep in mind that in many cases, you will need an API key, which you can usually get for free from museums. This key might be needed to keep you informed of updates, for potential account suspension in case of misuse, or simply for billing purposes in commercial offerings. Museums, luckily, provide their services as Open Data and typically offer them for free.

In R, it is quite simple. We define our endpoint and query, and then make the request:

```
library(httr)
endpoint <- "https://openaccess-api.clevelandart.org/api/
artworks/"
query <- "uchikawa"
call <- paste0(endpoint, "?q=", query)
get_artworks <- GET(url = call)
```

From the result, we can extract information like titles, image paths, and more. The result of the GET request (get_artworks) contains data in JSON format. To extract information such as titles or image paths, you need to parse the JSON content.

```
library(jsonlite)
content_data <- content(get_artworks, as = "text", encoding =
"UTF-8")
parsed_data <- fromJSON(content_data)
```

So, now that we know how to call an API in the cloud, let us explore using a cloud service for image recognition.

17.3 Vision APIs

Requesting a cloud service is essentially the same in principle. The only difference is that most of these services come with a cost, albeit a small one per request, but they are incredibly powerful, especially in the field of Computer Vision. Let us take a closer look.

> ℹ Note: What was previously known as "Computer Vision" based on Machine Learning is now generally referred to as "Artificial Intelligence" (AI). In most cases, it is the same technology; only the name has changed. The exception is generative AI, such as ChatGPT or DALL·E, which represent a technology genuinely new.

Let us start with a heads-up. The major players in Cloud Computing, namely Google, Microsoft (Azure), and Amazon (AWS), offer these services. The good news is that Microsoft Azure provides an R library that makes it incredibly easy to use. So, for this example, we will go with Azure.

First, you will need an Azure account, which you can initially use for free.[2] Later, it operates as a "Pay-As-You-Go subscription," meaning you only pay for what you use. You can monitor your costs and set budgets through the portal.

Within Azure, you can define a "resource" which will also be your endpoint. I named mine "your-resource," so my endpoint is https://your-resource.cognitiveservices.azure.com/.

You will also find an API key there (for your use only, do not share it).

Now, you will include both your endpoint and key in the first line of your R script:

```
library(AzureVision)


vis <- computervision_endpoint(url = "https://your-resource.
cognitiveservices.azure.com/",
                               key = "[your_key]")
```

17.4 Computer Vision via Cloud Computing

Let us have Azure analyze this image from the Cleveland Museum's collection (Fig. 17.3).

We provide this image via a URL:

```
image_url <- "https://piction.clevelandart.org/cma/ump.di?e=4985
0C7ACA000F14375725D3E7F324A7C10FE48EA335EBA44413087D66A53A7B&s
=24247294&se=1692237107&v=3&f=1996.412_o5.jpg"
```

So, what can we find in this image? Can the service describe it?

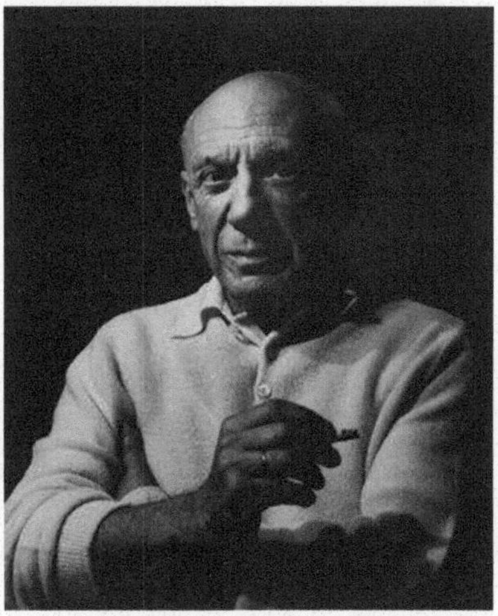

Fig. 17.3 Picasso, Cannes by Lucien Clergue, 1956 (Cleveland Museum of Art)

[2] https://azure.microsoft.com/en-us/products/ai-services.

To do this, we send both the endpoint and the image's path to the Azure service using the "describe" command, and here is what we get:

```
describe(vis, image_url)
```

```
$tags
 [1] "person"    "man"       "indoor"    "sitting"
"looking"   "table"
 [7] "holding"   "front"     "shirt"     "dark"      "food"
"glasses"
[13] "standing" "phone"     "young"     "laptop"

$captions
                          text confidence
1 a man looking at the camera  0.8803567
```

The response includes:

- Tags, which list various detected objects in the image.
- Captions, which provide a descriptive text for the image.

Interestingly, tags even include the term "Laptop." The list is sorted by how confident the system is in recognizing the object. It appears that the results after "dark" might be incorrect. You can also retrieve these values using the "tag" function.

```
tag(vis, image_url, language = "en")
```

```
             name confidence hint
1          person  0.9975610 <NA>
2      human face  0.9801561 <NA>
3             man  0.9762821 <NA>
4          indoor  0.9254409 <NA>
5 black and white  0.8968987 <NA>
6        portrait  0.7618407 <NA>
7        clothing  0.7530441 <NA>
8      monochrome  0.7483534 <NA>
```

But where exactly is this person located in the image?

```
detect_objects(vis, image_url)
```

```
  rectangle.x rectangle.y rectangle.w rectangle.h object
confidence
1           16          95          453         494
person        0.559
```

The recognized person is marked with a bounding box, indicating the upper-left corner (x and y) and its height and width (h and w).

You can also ask more broad questions. What category does the image content belong to?

```
analyze(vis, image_url)


$categories
     name      score
1 others_  0.00390625
2 people_  0.36328125
```

The categories in the result represent an estimation of what general categories the image content belongs to, based on pre-trained models. Each category is associated with a score, which is a probability value (ranging between 0 and 1) that indicates how confident the model is about its prediction. So probably it is a person. And we can determine the "area of interest," which is the most likely location of the main subject in the image:

```
area_of_interest(vis, image_url)
  x   y   w   h
  0  60 484 484
```

Here, you will receive the position and dimensions of the bounding box.

> ℹ Notice that this works well for photography, but less so for many art forms.

If you wish, you can also perform text recognition with this service.
read_text(vis, image_url, detect_orientation = TRUE, language = "en")

For precise information about the capabilities of this API, check out its description.[3]

[3] https://learn.microsoft.com/en-us/azure/ai-services/computer-vision/.

17.5 Another Example of Cloud Computing

I created a Mastodon bot for art. Mastodon is a decentralized microblogging network like Twitter. My account is named `@HiddenGems`, and it displays objects that are not on display in museums but are found in their digitized collections and are rarely seen.

Four times a day, the bot fetches an artwork from the Cleveland Museum of Art (and the Rijksmuseum), describes it in an engaging way, and posts it using various APIs. Here is how it works:

1. I use a weather API to check today's weather in Cleveland and receive a JSON file with a brief weather description, like "Clouds."
2. I pass this string as a search term (and use "has_image=1") to the Cleveland Museum's API, which returns a JSON file with several artworks that contain the search term in any field. For "Clouds," I get images of clouds as well as images from "St. Cloud" near Paris, which is also quite interesting.
3. From the list of artworks, I randomly select one and send its metadata, "wall_description," and "fun_facts," along with a prompt to OpenAI's API. In return, I receive a text for posting.
4. I then download the image from the Cleveland API, send it along with the text from OpenAI, and include a link to the corresponding webpage showing the artwork (from the Cleveland API) to my Mastodon server's API to make the post.

Here is an example of what it looks like (Fig. 17.4):

The entire process and its daily execution can be orchestrated on a server using R triggered as a cron job. A cron job is a scheduled task in Unix-based systems that runs automatically at specified intervals, such as daily or hourly, to automate repetitive processes. However, I have now moved it to a cloud-based No-code platform.

> ℹ If you are on Mastodon, feel free to follow the account. If you like it, recommend it to others.

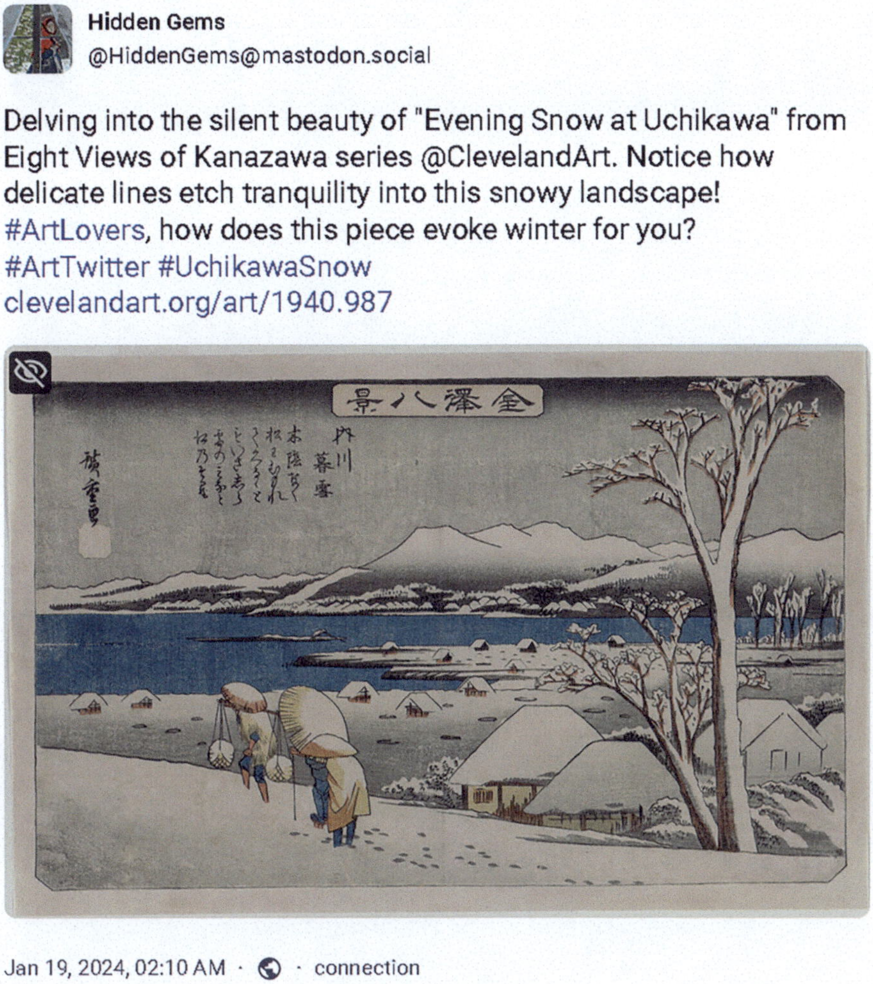

Fig. 17.4 Automated post on social media platform Maston

Chapter 18
Epilog: The Future of Cultural Data Science

For those looking to delve deeper into cultural data science, there are numerous resources available. The R programming language is a powerful tool for data analysis and visualization that could be given only an introduction to. The comprehensive R archive network (CRAN) offers extensive documentation, which is an invaluable resource for both beginners and experienced practitioners alike. Additionally, platforms such as R-bloggers and Stack Overflow provide a wealth of knowledge and community support for troubleshooting and refining your code. Conducting a Google search will reveal a vast array of resources and support.

CRAN (Comprehensive R Archive Network) CRAN is the primary repository for R packages and documentation. It offers detailed manuals, FAQs, and a wide range of contributed packages that extend R's functionality. Whether you need to perform basic data manipulation or complex statistical modeling, CRAN is your go-to source for official documentation and user-contributed packages.

https://cran.r-project.org/

R-bloggers R-bloggers is a popular community-driven platform that aggregates content from hundreds of R blogs around the world. It is an excellent resource for tutorials, tips, and tricks from other R users. The content covers a wide range of topics, from basic R programming to advanced data visualization, machine learning, and statistical analysis.

https://www.r-bloggers.com/

Stack Overflow Stack Overflow is one of the largest online communities for coding. The R tag on Stack Overflow is an invaluable resource for troubleshooting code issues, seeking advice on best practices, and finding code snippets for specific tasks. The community is active, and you can usually find answers to most R-related questions or ask your own.

https://stackoverflow.com/questions/tagged/r

H. Klinke, *Cultural Data Science*, Quantitative Methods in the Humanities and Social Sciences, https://doi.org/10.1007/978-3-031-88130-5_18

Coursera Coursera offers a variety of online courses on R programming. These courses range from introductory to advanced levels and are often taught by professors from top universities. They cover various aspects of R, including data science, statistics, and machine learning, with a mix of video lectures, readings, and hands-on exercises.

https://www.coursera.org/courses?query=r%20programming

In recent years, the rise of artificial intelligence (AI) and large language models (LLMs), like ChatGPT, has revolutionized the way we approach coding and problem-solving. These tools are not only capable of generating code but also of offering insightful solutions to complex error messages and suggesting ways to optimize code. Integrations like GitHub Copilot in RStudio or ChatGPT coding assistant exemplify how AI can streamline the coding process, making rapid prototyping more accessible and efficient.

More complex tasks, such as advanced machine learning, web development, or the deployment of code to a server, might require the use of another programming language like Python, which offers broader libraries and frameworks tailored for these applications. Furthermore, advancements in artificial intelligence, such as the paid version of ChatGPT (Pro), now allow users to upload data and interpret code directly within the chat interface. This capability makes simple data analysis incredibly straightforward, enabling users to quickly perform calculations, generate visualizations, and derive insights without needing to leave the coding window.

By leveraging these resources, you can continuously improve your R programming skills and stay updated on the latest developments in the field. Whether you are working in cultural data science or any other data-driven discipline, these tools and communities will help you unlock the full potential of R.

As we conclude this exploration of cultural data science, it is important to reflect on the rapid advancements in technology and the significant opportunities they present for future research in this field. Looking ahead, the future of cultural data science is likely to be shaped by several key trends and developments.

LLMs have begun to make significant inroads into cultural data science. These models can generate text, summarize content, and even provide interpretative insights into cultural data. Their application extends beyond textual analysis to include multimodal data. Multimodal data refers to datasets that combine different types of information, such as images, text, audio, or video, enabling richer analysis by leveraging the relationships between these diverse forms of data. Here, LLMs can assist in interpreting visual and textual data together.

In the context of cultural data science, LLMs offer the potential to bridge the gap between quantitative data analysis and qualitative interpretation. For example, LLMs can be used to generate narratives or contextual explanations for the patterns detected in cultural datasets. This ability to combine data-driven insights with human-like interpretation makes LLMs a powerful tool for cultural data science.

Moreover, LLMs can assist in the rapid prototyping of research ideas by providing code snippets, debugging assistance, and suggestions for data analysis techniques as mentioned above. This capability enables researchers to experiment with new methods and approaches more efficiently, accelerating the pace of innovation in the field.

As cultural data science advances, it inevitably raises important ethical questions. Particularly, bias is an increasingly recognized challenge, especially in the context of machine learning and computer vision systems. These systems, as discussed by Offert and Bell, often inherit biases not only from the datasets on which they are trained but also from the inherent structures and algorithms used in their development. This "perceptual bias" can lead to skewed interpretations of cultural data, particularly in fields such as digital art history, where the nuances of visual culture are critical [1].

For instance, when machine vision systems analyze artworks, they may prioritize certain stylistic elements or iconographic motifs over others, based on the bias embedded within their training data. This can result in a homogenized view of art history, where diverse artistic expressions are underrepresented or misinterpreted. Such biases are particularly problematic in the study of non-Western art, where the datasets used to train these models may not fully capture the diversity and richness of global artistic traditions.

The use of such advanced techniques also raises important methodological questions. Impett and Offert argue that the application of these models requires a new critical methodology that considers the epistemic entanglement between the models and the data they analyze [2]. This means that any analysis conducted using these models must also include a critique of the models themselves, recognizing that the insights they produce are shaped by the training data and the inherent biases of the models.

Rodríguez-Ortega emphasizes the need for spaces of convergence and collaboration, where mutual feedback between humanists and technologists can lead to a better understanding of both human culture and the technologies we use to study it [3]. By fostering a dialog between different disciplines, we can ensure that the technologies we develop are not only scientifically rigorous but also culturally informed and ethically sound. The future of cultural data science, therefore, lies in the intersection of data-driven methodologies and humanistic inquiry. Art and culture are not merely collections of data points to be quantified—they are expressions of human experience, shaped by complex social, historical, and emotional factors. Culture cannot be fully understood through quantitative methods alone. As such, the future of cultural data science must be one that embraces both the empirical rigor of the sciences and the interpretive depth of the humanities.

This is a beginning of new questions, new explorations, and new understandings in the ever-evolving field of cultural data science.

References

1. Offert F, Bell P. Perceptual bias and technical metapictures: critical machine vision as a humanities challenge. AI Soc. 2021;36(4):1133–44. https://doi.org/10.1007/s00146-020-01058-z.
2. Impett L, Offert F. There is a digital art history. Vis Resour. 2024;38:186–209. https://doi.org/10.1080/01973762.2024.2362466.
3. Rodríguez-Ortega N. Image processing and computer vision in the field of art history. In: Brown K, editor. The Routledge companion to digital humanities and art history. New York: Routledge; 2020. p. 338–57. https://www.taylorfrancis.com/chapters/edit/10.4324/9780429505188-30/image-processing-computer-vision-field-art-history-nuria-rodr%C3%ADguez-ortega.

Index

H. Klinke, *Cultural Data Science*, Quantitative Methods in the Humanities and Social Sciences, https://doi.org/10.1007/978-3-031-88130-5

Q

R

S

T

U

V

W

The manufacturer's authorised representative in the EU is Springer
Nature Customer Service Centre GmbH, Europaplatz 3, 69115 Heidelberg,
Germany. If you have any concerns regarding our products, please
contact ProductSafety@springernature.com

Printed and bound by CPI Group (UK) Ltd, Croydon, CR0 4YY
30/09/2026
02225577-0014